Earthdream

The Marriage of Reason and Intuition

Earthdream
The Marriage of Reason and Intuition

ROBERT HAMILTON

First published in 1990 by
Green Books
Ford House, Hartland
Bideford, Devon EX39 6EE

© Robert Hamilton 1990

Typeset in 11 on 13 point Erhardt
by Fine Line Publishing Services, Witney, Oxon

Printed on recycled paper

British Library Cataloguing in Publication Data
Hamilton, Robert
Earthdream.
1. Ecology
I. Title
574.5

ISBN 1-870098-11-0

Grateful acknowledgement is extended to the following for permission to
reprint copyright material: Cambridge University Press for the excerpt from
My View of the World by Erwin Schrödinger; Laurence Pollinger Ltd and the
estate of Frieda Lawrence Ravagli for lines from *Etruscan Cypresses* by
D.H. Lawrence; Harper & Row for Chapter 1 and other extracts from the
translation of the *Tao Te Ching* by Stephen Mitchell; Wildwood House Ltd
for Chapter 14 from the translation of the *Tao Te Ching* by Gia-fu Feng and
Jane English.

Contents

to the memory of my father, Ken

Acknowledgements

THIS BOOK is built upon the pioneering work of a number of great writers and visionary thinkers, and it is to these original minds that I owe my greatest debt. Perhaps the best way I can repay them is by pointing you to their work, so that you can retrace for yourself the path that has led me to the synthesis of ideas presented here. In no particular order, then, I would like to thank the following for the inspiration which has brought this book into being: Richard Bach, Erich Fromm, Gary Zukav, Arthur Eddington, James Jeans, Erwin Schrödinger, Albert Einstein, Fritjof Capra, Roger Jones, Robertson Davies, Rudy Rucker, Fritz Schumacher, Robert Pirsig, Gregory Bateson, Morris Berman, Paul Davies, Carl Jung, Ken Wilber, Colin Wilson, Joseph Campbell, James Carse, Alan Watts, Kahlil Gibran, Marilyn Ferguson, Abraham Maslow, Rollo May, Laurens van der Post, Douglas Hofstadter, Thomas Berry, Satish Kumar, Jonathon Porritt, Kathleen Raine, James Lovelock and, finally, although far from exclusively, Lao Tzu.

I feel very fortunate to have been able to take the time out from 'real' work to research and write this book. It would never have been possible without the financial help of my mother, Dorothy, who has never failed to support me, in whatever madcap scheme I have come up with—and I am sure she thought that setting out to write this book was the maddest of them all! As well as my mother, I am grateful to a number of other people for housing and feeding me during my sabbatical, most particularly Mavis Flaherty in Larchmont, Roy and Bev Gower in Te Miro, and Dave and Chris Gower-Rudman in Takanini. Many thanks are also due to my sister, Jan, for her help with typing in the early stages, and to Jim Flaherty for his help in keeping the wolf from the door during the final stages of writing and editing. On a very different front, I would like to express my special thanks to Dave Gower-Rudman, Graham Moxham, Frank Thomas, and Peter and Sarah Haines for their help in taking my mind away from the writing of this book—by getting me out running on the hills. Their companionship has been greatly cherished.

The most special thanks of all, though, must be reserved for my wife, Shannon O'Flaherty. Her keen-eyed editing always found me out if I was being lazy with my use of words, or steered me back on course if I was drifting away on some tangent, but, more than anything else, her love and encouragement and faith have given me—sometimes against quite stubborn resistance—the self-belief that I have needed in order to keep my commitment charged. *Earthdream* belongs very much to the two of us.

R.G.H.
Ilkley
August 1990

Prologue

THE ORIGIN of this book can be traced directly back to one particular moment of experience in the summer of 1985. I was in California for a month's holiday, and spent some time in Yosemite National Park. A visit to this very special place represented the fulfilment of a long-held ambition. Yosemite, just the name alone, had always held a certain magic for me. And here at last I found myself. There was no sense of anti-climax. The real Yosemite was every bit as magical as the Yosemite of my imagination. After a little time spent exploring the trails from a base in the valley, during which I duly made my pilgrimages to the tops of its two most famous peaks, El Capitan and Half Dome, I was set, this one particular day, to go off hiking into the back country. Although I had previously spent hundreds of days walking, running and climbing in the mountains, this was the first time I had ever freed myself so completely from logistical constraints. I had no fixed route, no fixed schedule, no need to get to any specified place at any specified time—at least not until my food ran out!

As I climbed out of Yosemite Valley, leaving the swarming tourists far behind, it was as if I were also climbing out of time. I could not ever remember feeling such an exquisite sense of peace and tranquillity. The daily pressures of twentieth-century living were left so far behind that instead of just being forgotten, as usually happens in the hills, they were perceived to be utterly insignificant. They just melted away into the landscape. As I settled into the rhythm of walking, pack comfortable, pace relaxed, weather perfect, I became more and more lost to my surroundings, feeling, by way of some strange but sublime synthesis of object and subject, that I was being absorbed into the trail and the trees and the talus. I

I

was filled with a serene joy, and a richness of *being* that made the Western idol of *having* appear simply irrelevant.

At the end of this first idyllic day's hiking in the wilderness of the backwoods, staring around the small clearing in which I had camped, I found myself being enveloped by the silence of my solitude, inducing in me some strangely ambivalent emotions. A sense of security was mingled with a sense of vulnerability. A sense of community was mingled with a sense of isolation. There was both a felt empathy with the forest and a felt detachment. Although I was at ease in the little clearing I had found and was feeling rather proud of my self-sufficiency, I was also aware of just how reliant I was upon my hi-tech lightweight equipment. I felt close to nature, yet at the same time I was being reminded just how far I was actually removed from nature. I felt both a sense of belonging and a sense of not belonging. Then, just as I was beginning to absorb and explore this contradiction, my reverie was suddenly interrupted. Out of the corner of my eye, I spotted what at very first sight appeared to be a bumble-bee gathering nectar. An instant later, I realized that it was actually a humming-bird.

Never having seen one before, I was immediately struck by its diminutive size and extraordinary fragility, but then, almost in the same instant of recognition, something else hit me, deep inside. My very being, it seemed, had become seized by the humming-bird's absolute perfection. I had never before experienced anything quite like the feeling that flowed through me in that moment. My day of easy, solitary, timeless walking had somehow tuned me in. The landscape had made me receptive. It contained a sense of undiluted astonishment at the sheer fact of existence. It contained a sense of deep spiritual affinity with the humming-bird. It contained a sense of profound connectivity, an awareness of kinship, that we shared the same ineffable origin. It hinted at connections that exist beyond human comprehension. It was a moment of wholeness. An epiphany. Numinous. In a way that just cannot be rationalized, I was granted a sense of the sacred. It recalled to mind childhood memories of gazing into a star-studded night sky, memories filled with shades of emotion that had long been forgotten; feelings of great reverence, of innocent wonder and boundless curiosity before the immensity of the universe, those same feelings of infinitely deep mystery that have stirred people's imagination since the very beginning of human history. It was as if I were seeing physically through the unprejudiced eyes of a child, but mentally through the inner eye of some very ancient wisdom.

The humming-bird, perfectly innocent of the fact that it was the cause of such an eruption of emotion, shied away from my avid attention and flew off, disappearing into the depths of the forest. And somehow, that seemed right. The mysteriousness of the experience was preserved. The enchantment, the spell under which I was cast by the little creature's spontaneous appearance, was left unbroken. And it remains unbroken to this day. I am *still* enchanted.

The humming-bird gone, the sound of natural silence restored, I was absorbed once more into the restful ambience of the backwoods. Now, though, my emotions were far less ambivalent. I felt a very assured sense of belonging, and an incredible, rippleless calm of mind—what was to be the calm before the storm. Standing there, quite alone, meditating, it suddenly became impossible to suppress a huge grin, the afterglow feeling of joy and meaning becoming intensified into a moment of startling realization. I underwent a conversion experience. I wasn't sure, at that point, to exactly *what* I had been converted, but I could no longer label myself a strict scientific rationalist. In a dramatic slippage of consciousness, a kind of cascading mental avalanche of excitement and bewilderment, the whole logical way in which I had always looked at the world suddenly fell apart.

During the next four or five days' walking I was never quite able to recapture that same sense of calm and belonging I felt on that wonderful first day. I suppose my mind was too absorbed analytically to tune into the environment around me, too taken up with assimilating the meaning of the events of that first night in the forest. Questions were breeding rapidly in the hotbed of a suddenly restless and fertile mind. They gave me no peace. After many years of hibernation, the insistent curiosity of my youth had been violently disturbed from its slumber. My thoughts were again led back to those childhood memories of gazing into the black infinity of space, to those first mystical encounters with the overwhelming inexplicability of the universe. How was it that I ever lost that innocent sense of wonder? How was it that I ever lost that sense of the enchantment of the world? I did not really know, and it did not really matter. I just felt grateful for the reawakening of my spiritual senses. I seemed to be reliving the tremendous excitement that I felt as a young teenager when I discovered the world of mathematics for the first time, experiencing the same anticipation-filled tingle, like an explorer who is just about to set foot on a new and uncharted continent.

I realized that something had been drastically missing from my education, that because of my scientific training and mathematical bias, an entire epistemological dimension had been completely neglected. In sixteen years of school, college and university, I had never once seriously taken account of value and meaning. Virtually all my study had been carried out in a kind of vacuum, in the abstract, from a totally objective viewpoint, looking upon the world as an external reality, completely independent of me as observer. The human mind had no role to play in any discipline that I had ever studied. The dogma was that the mind was not needed in the working description of the universe as provided by science. It was not necessary. Mind was just an epiphenomenon. This was the rationality that I had grown up with at school and college and was never in any way led to doubt—until now. My forest epiphany had allowed me a glimpse through this ideology, opening up a whole new vista of intriguing psychological, philosophical and epistemo-logical questions. I began to hear the whisper of an inner voice urging me to step off society's conveyer-belt and take some time to try and understand this challenge to scientific rationality which had just been presented to me. I wanted to search for a deeper kind of rationality, a conceptual framework in which I could possibly understand and validate the qualities of my ennobling 'conversion' experience.

Following my mountain soliloquy, this intoxication with the deeper mysteries of life inevitably led me to ponder over the *problems* of life. I had never before been quite so deeply aware of the rift which has occurred between the designs of humanity and the patterns of life on Earth, nor so deeply troubled by the frightening extent to which that rift is being forged wider as technology becomes an ever more dangerous tool in our greedy hands. The vast power granted to us by our modern technology carries with it a vast responsibility. I began to feel some discomfort at being a participating member of a society which was so arrogantly disinclined to accept that responsibility, a society so totally unwilling to pay the environmental price that is inevitably tagged on the back of our high material standard of living. It was these very concerns that were buffeting around in my head as I travelled back from Yosemite toward San Jose, having hitched a lift with two climbers I had met earlier while descending the small glacier below the summit of Mt Lyell—the finale of my hiking trip. Indeed, perhaps it was just this dramatic contrast that had provoked my unease: standing in seemingly timeless solitude atop a 14,000-foot mountain in the morning; travelling on a busy highway in a conspicuously twentieth-century setting in the afternoon. The rapid change in perspective had served to focus my thoughts.

We had just left the flat expanse of the San Joaquin Valley and had started to climb into the rolling, sun-parched grasslands of the Diablos when I was again unexpectedly disturbed from a quiet meditation. The natural sweep of the contours was interrupted in a quite extraordinary way. In seconds, we had been transported from a pastoral to a futuristic landscape. The horizon was now dominated by the hard lines of technological machinery—although strangely not in any kind of obtrusive way. Despite the fact that I had woken up that morning to the unspoilt scenery of the High Sierras and was still very much tuned in on nature's wavelength, this artificial industrial landscape did not jar upon my sensibilities. I was astonished to find myself regarding it with the same reverent awe as I had that mountain wilderness a few hours earlier. My acceptance of this intrusion rested upon the fact that the technology was entirely appropriate. In fact, the machinery belonged to a very unusual farm, an energy farm, harvesting the wind. The hills were alive with the motion of hundreds of windmills, the space-age variety, streamlined towers with aerodynamic rotor blades, each turning slowly, rhythmically, almost as if they were choreographed, producing electricity from the wind, generating power in partnership with nature, without consuming resources, without creating waste. It was astonishing. It was surreal. It seemed to offer a vision of the future. I began to dream.

A few days later, I arrived back in England with a troublesome companion—a vivified conscience which refused to let me sit back and evade the difficult questions that had just been brought to the surface. I had become sickened by the apathy and complacency of the society in which I lived. But I had also become aware that society's apathy embodied *my* apathy, for none of us can truly be exempted from a share of the responsibility. I had to own up to my own share. And it was something I found very uncomfortable. I realized that I could no longer hide in the crowd and turn the conventional blind eye. I could no longer hide from myself in the secure habituality of normal, everyday boredom. That inner voice had now grown from a whisper to a plaintive cry. It was insisting that I go ahead and search for that deeper rationality. Suddenly growing aware of just how destitute I had become of the sense of free will, I determined that I would no longer settle for having my choices selected for me by the external circumstances and events of my life. I determined that I had to escape the spurious security of my boredom. I set out to pursue my dream.

This book is the result, although not, by a very long way, the fulfilment of that dream. My dream is a very big dream, a very great dream, and it will require a very great many people to share in it before it could possibly be realized. I have given this dream a name. Inspired by Thomas Berry's poetic and moving book, *The Dream of the Earth*, I call it the *Earthdream*. Taken as a whole, then, this book is written in the hope that it might contribute to the individual understanding that will be needed to seed this dream. And it is dedicated at large to those readers who have the vision and courage to share in and work toward its realization—to all those fellow Earthdreamers out there. Pursue this dream with me.

Introduction

E ACH ONE OF US, as a human individual, is very much the product of the way in which we have come to perceive ourselves and the world about us. The way we think and speak and behave is conditioned by our fundamental beliefs about what it means to be a human being. To a considerable extent we come to be defined by our own self-definition, by the way we look at ourselves and our relationship to the world outside of ourselves, in other words by our own personal mythologies, our belief systems, our metaphysical maps. The trouble is that we do not enter the world with such a map. Having unshackled ourselves from instinctive determination, each one of us has to construct our own frame of reference from which to try and make sense of the world, to find a position from which to argue the decision-making process of life.

In the past, the innate psychological need for some such metaphysical map has always been satisfied through the automatic inheritance of the map of the culture into which each man or woman was born, a map which, up until the age of technology, has invariably been explicitly defined. In the modern era, though, and quite uniquely, this process is no longer straightforward. The animating mythology of today's world is not well defined, and remains largely unspoken. Too disturbing and uncomfortable to be openly acknowledged, it leads a kind of surreptitious existence beneath the Western consciousness. It implicitly provides the rationality which underlies the whole organization of contemporary society, but, at the same time, is largely disowned by that society.

The confusion can most clearly be witnessed in education where we find that there is no general scheme of presentation through which children have an opportunity to develop an authentic set of beliefs about the world and a coherent sense of who they are as human individuals. In fact, such is the sensitivity to the subject of metaphysical cartography that schools seem to actually suppress our children's sense of inquiry into the world. Students can emerge from fifteen or more years of intensive study without ever having had to turn their perception inwards to face and question the miracle of their own existence.

In the West, for most of our recorded history, it has been systematized religion that has provided the mythology by which people have lived. Supposedly mandated by God, the Christian Church has functioned as a lighthouse on the sea of life, illuminating the way, revealing clear landmarks, demarcating good and evil, moral and immoral, defining sin and offering salvation. It has offered an explicit, unambiguous metaphysical map with quite precise directions. However, during the last 300 years, against the irresistible, relentlessly rising tide of scientific knowledge, its guiding beam has gradually been dimmed to the extent that now orthodox religious belief plays little authentic role in the life of society at large. This decline in the role of traditional religion in human life is the single most important fact of our entire history. In a way, it represents a sign of increasing maturity, marking our passage through an inevitable stage in our slow growth toward spiritual adulthood. In its collective social and cultural evolution, Western humanity is now in its adolescence, and as the metaphor might suggest, it is a difficult stage of growth. The fact is that although some 300 years have passed since the birth of the scientific revolution, the Western consciousness has yet to come fully to terms with its loss of external religious value. It is still experiencing adolescent growing pains, still lacking the spiritual maturity needed to embrace an adult mythology.

Essentially, a mythology stands as a particular way of understanding the meaning of reality and human existence. A mythology is a collection of beliefs which has evolved to meet our deeply-rooted need to have some shape brought to our experience of life, to have some pattern imposed upon our perception of the world. A mythology is a cultural response to the human situation, as

8

shaped by a particular historical, geographical and social context. If that context changes and the prevailing mythology becomes inadequate to the needs of society, tensions are created which will work toward its transformation. This is the situation in which we find ourselves today. Recognizing the limits to growth, recognizing the interconnectedness of our environmental and social problems, recognizing that, ecologically and spiritually we exist as one global community, our context has changed. Our mythology is no longer appropriate to our needs. The forces of change are building now, much as they did during the birth of the scientific revolution in the seventeenth century.

That last major cultural transformation marked our point of departure from spiritual childhood into adolescence. For the pre-scientific mythology of the Middle Ages was indeed characterized by a childlike innocence and a dependence upon the mother—as represented by the Church. The medieval world was an enchanted world, a world suffused with meaning and purpose. Its mythology—a union of Christian theology and Greek philosophy—was a very substantial reality in the life of every individual, providing a highly rational, pre-ordained order behind each element of the human, natural and heavenly realms. Human existence was considered to be tangibly connected to a transcendent level of being. The social order was deemed to have been conferred by God with divine grace passing from generation to generation through the blood, ensuring that the feudal system of those times was not only self-legitimating but also self-perpetuating, the life of each individual being rigidly constrained by its inherited position within the social hierarchy.

But this oppressive social and cultural constraint was also the source of a level of existential security which many would probably envy today. The psychological life of humanity in medieval times was hermetically sealed. The symbols, dogmas and rites of the Christian religion provided for a highly structured, undisturbed passage through life. There was little act of will required, and therefore little felt anxiety. Recognized unequivocally as God's intermediary authority on Earth, the divinely sanctioned arbiters of his mythic symbols, the Church held a position of unrivalled supremacy. Functioning as nothing less than a religious dictator-ship, it ruled over all social and cultural affairs. It provided both the physical and the spiritual fabric for a society which survived for many hundreds of years in a state of equilibrium, albeit an inherently unstable one. Just as it is inevitable that a child will come to lose its innocence, so it was always inevitable that the foundations of medieval society were going to be undermined.

The two principal changes in context which eventually precipi-tated this collapse were provided by the development of classical science and the failure in the continuing economic viability of the feudal system. The rationality that emerged from the ensuing mythological transformation found its avatar in René Descartes, although it should be understood that the ideas and concepts that he embodied belong far more to the milieu of his time—the first half of the seventeenth century—than to any one individual personality. The new rationality was very much 'in the air'. It was simply Descartes who provided its most succinct and complete expression. By accepting nothing into his system unless its truth was evident with mathematical certainty to his doubting reason, he attempted to build a philosophy of nature from very first principles, and in the process became the chief instigator of the analytic method: reducing complex problems or structures to their con-stituent parts and then analysing the whole by a logical ordering of these smaller units. A complete understanding of the whole could be gained from an understanding of the component parts, which, when reduced to sufficiently simple forms, could be perceived directly and indubitably with the mind. This was the key which would eventually give the human mind access to all knowledge.

Supported by his mathematical discovery of the beautiful correlation of algebra and geometry, which suggested that all of physics could be described in geometrical terms, Descartes became convinced that the material world was to be understood fully by simple mathematical reason. Rigidly compartmentalizing the world of the senses and the world of matter, he reinforced the dichotomy which was already implicit within the old Christian world-view: the dichotomy of mind and body. The mind, as subject, confronted an external world as separate object. This absolute partition allowed all the spiritual and aesthetic, sensory qualities of life to be placed neatly into the one compartment, leaving the other free to be expressed completely and perfectly by way of mathematical law. He saw the material universe purely as a machine, with the explanation of all physical phenomena lying simply within the motion of its component parts.

With the art of clock-making reaching its zenith during this period, a natural symbol for the new world order was at hand: the clock, being held to represent mechanical perfection, provided the perfect metaphor for the universe. The material world was seen to function automatically, ticking away, blindly executing a precise programme through the mechanics of its inner workings. However, this vision of the world as God's automaton could remain only a vision in Descartes' lifetime. It was destined to be realized by the greatest intellectual figure of the Renaissance, Isaac Newton, whose monumental achievement was a universal formulation of the laws of motion and gravity.

Before the publication of Newton's *Principia* in 1686, Kepler's empirical laws describing the motion of the planets and Galileo's law for falling bodies were seen to belong to distinctly different realms: Kepler's to the celestial and Galileo's to the terrestrial. God had ordained that heavenly bodies move in circles and earthly bodies move in straight lines. Newton performed a synthesis of these two laws. He saw that the motion of an apple falling to the ground and the motion of the Earth around the sun were dictated by the same natural principle—gravity. This discovery represented an incredibly creative insight, the scope of which is impossible for us to put into proper perspective today. Quite unexpectedly, and almost by accident, Newton had cracked the most fundamental code of nature—or so it seemed. Just as significantly, he also managed to frame his universal laws of motion in a new and extraordinarily powerful mathematical language called the differential calculus. In this framework, the difficult concepts of instantaneous rates of change of position and velocity were handled with beautiful simplicity. He had reworked Galileo's mechanics in a more complete way and at the same time had managed to incorporate his own principle of gravity, and all in a remarkably concise new language. His laws were so supremely elegant in expression that no one could possibly doubt their truth. Newton was able to give a complete mathematical, geometrical formulation to Descartes' vision of the universe as a perfect machine. The mechanistic, atomistic, deterministic world-view which was to dominate the future development of all branches of science had been firmly established.

In the span of just a single lifetime, the way in which reality was perceived had been completely transformed. Instead of being envisaged in human, *subjective* terms, the universe was now looked at principally in scientific, *objective* terms, from a level removed from that of the natural world, without reference to any human observer. Instead of thinking about *why* things happened in nature, it was now only important to describe *how*. Instead of being looked at as a whole, holistically, with explanation sought in terms of relationships to a higher order, the universe was now considered to be split into two, with the material side perceived atomistically, as a vast collection of small, independent, mechanically interacting parts. However, despite these very great modifications in perception, the new pretender had not yet overthrown the reigning Christian mythology. Science was still contained within the greater context of religion. By enabling nature to be abstracted as a world of manipulable quantity, divorced from the world of quality, the new rationality came to be regarded as a God-given instrument of control. Seemingly with divine blessing, science promised to make humanity master of its terrestrial environment, a prospect which was to prove irresistibly attractive to the leading intellectuals of the time. Blinded by the glittering illusion of ever-increasing returns, the metaphysical flimsiness and embedded problems of Descartes' dualistic understanding of reality—like, for example, the lack of any interface between mind and body—were simply overlooked and were set to remain so for more than 200 years.

Along with the new rational ordering of the natural world came a new, increasingly abstract, rational ordering of human social and economic relations. The birth of the modern scientific spirit was accompanied by the birth of the capitalist spirit. For a long while, though, these new patterns of perception continued to develop within the embrace of the pattern of Christianity. Following the scientific revolution, throughout the eighteenth century and into the nineteenth, meaning still firmly rested with God—although in a somewhat modified way. God as the omnipotent creator was now held to be the Celestial Engineer of the great universe machine, and as such had been relegated from a dynamic role to a rather more passive one. He was now only required to set the universe in motion—to wind up the clock—after which His attendance was no longer strictly needed. It was not necessary for Him to oversee its daily mechanical workings. God was now a step removed from reality, and as the scientific and capitalist spirit gradually gathered momentum, so this distancing process continued, eventually culminating in 1859 with the publication of Charles Darwin's *Origin of Species*, when the stage was set for science finally to banish God altogether.

Very simply, Darwin's theory purported to show that the plant and animal kingdoms, including humanity, were to be regarded as a product of natural selection—survival of the fittest—working on hereditary variation. Although his theory is rather crude in comparison with its modern derivative, its simple beauty and tremendous explanatory power totally captivated the scientific community, especially when backed up by the vast wealth of supporting evidence which Darwin had accumulated. Essentially, he succeeded in removing God yet further from the scene of action by arguing against the need for design: the living world had *evolved* from very simple beginnings. The guiding hand of God was no longer required. All that was really deemed necessary was just his signature on the original proposal. The rest looked after itself. Science was at last able to break completely free from the religious context, an autonomous mythology defining its own context.

The trouble was that by demolishing the metaphysical reality of God, science had effectively demolished its own metaphysical foundations, the reasoning behind Descartes' dualism depending upon the fact of the existence of God for its validity. It was Descartes' absolute faith in God that underpinned his absolute faith in the power of human reason. But science had now cut itself off from this support, and again dazzled by the glittering prizes that seemed to be promised, the intellectual community was blinded to the fact that something was seriously wrong.

The mechanism of Darwinism fitted perfectly into the well-oiled cogs of the nineteenth-century universe machine, providing the justification for the same reductionist techniques that had been so successful in physics to be applied, quite uncritically, to the subjective sciences. Most indefensibly, the Darwinian principles of competition and survival of the fittest were taken on board and used to defend the wretched human exploitation which had fuelled the industrial revolution. Instead of the social order being conferred by God, it was now deemed to have been conferred by the natural forces of evolution. The privileged position enjoyed by the upper-class white élite was vindicated precisely by virtue of its privilege. Their supremacy was held to be self-validating. The world of human affairs came to be viewed in the same way as the natural world was now viewed—as a world of conflict: different classes, different races, different nations, competing against each other in struggles for power, struggles which were seen to represent the fundamental driving force of progress. Rather ambiguously, this fragmented picture of the world provided a theoretical framework which could not only be used to justify and reinforce the ideology of capitalism, but also to criticize and dismantle it, as Karl Marx first did so brilliantly with *Das Kapital*, the launching pad of communism, capitalism's terrible younger ideological sibling. However, of more importance than the virtues of either set of arguments was the fact that, subsequently, using such a framework, all leading social and economic philosophy came to be characterized by an appalling lack of respect for human individuality. The supremacy of industrialism passed beyond dispute, the individual lost as a mere bit-player in the corporate struggle. The unspoken mythology by which we live today had arrived at its logical completion.

Put in other words, by the end of the last century the process of abstraction represented by the move from qualitative to quantitative value had gathered to itself the unstoppable momentum which has brought us to our present position, where all the more subtle values of life have become subordinated to the abstract economic values of the market—the free market in the West and the state market in the East—and where, in both East and West, the cynical manipulation of public taste and demand and opinion is accepted virtually without protest. The divorce of production from almost all spiritual or environmental considerations has been fully legitimated in the sacred name of efficiency and profitability: the creed of the capitalist spirit. Owing allegiance only to the mathematical laws of economics, the supreme rule of the profit motive has granted industry free licence to compete with nature on whatever terms its technology allows it to dictate. The natural landscape is to be subjugated. But, as I have already intimated, applying this process of abstraction to nature has inevitably meant that it has also been applied to humanity. As nature has become despiritualized, so the lives of individual people have become despiritualized. As nature has become disenchanted, so the lives of individual people have become disenchanted. Western humanity, the youthful adolescent, finds itself both orphaned and homeless.

At the centre of this mythology is the modern theory of evolution, which sees human beings as having developed quite by chance out of the building blocks, the organic chemicals, emerging from the primordial sludge existing on Earth some 4,000 million years ago. We are told that our precious planet is but an insignificant speck of dust in an indescribably vast universe, revolving around an insignificant star among billions in our galaxy, itself insignificant among billions of other galaxies. All matter is constituted of elementary particles of inert matter and energy, the interactions of which can explain all the processes of the universe from the heat of the sun to the function of the human mind. Life on Earth arose by sheer fluke when self-replicating molecules spontaneously appeared as a result of some fantastically fortuitous freak of circumstance, and then, by an even more fantastically fortuitous freak of circumstance, managed to organize themselves into single-celled organisms. From these humble beginnings, by a process of random copying errors and natural selection, *you* are here today reading these words.

The implicit message of fundamentalist science is that we are absolutely insignificant in a cosmos which is totally indifferent to our aspirations and aesthetic values. We are to be seen at best as an intelligent robot, and at worst, as some insignificant cog in the mindless universe machine. We have no destiny, and no duty. Our icy solitude in this cold, soulless, infinitely apathetic space is total and unbridgeable. There is no spiritual rainbow. We are stranded with no identity outside of the individual ego. Indeed, the only personal value that can be validated by today's scientific fundamentalism is that of self-interest. Here, in this mythology of disenchantment, is to be found the generating cause of the boredom which has so torpefied the modern Western imagination. Here is the source of the insidious apathy and complacency, the cowardice, which surely lies at the bottom of so many of our current social and environmental problems.

Nevertheless, in spite of these grave mythic implications, it has to be said that the gradual unravelling of the mysterious forces at work in our universe has been *the* supreme achievement of human intellect and endeavour. In this stunningly successful quest for rational explanation, science has brought together men and women from every corner of the Earth in one common striving. It has provided a unique unifying force in the world. But tragically the scientific establishment has largely been unable to communicate its magnificent achievements and their *real* meaning to the people of the world. Science is generally just as remote and inaccessible as religion. Both institutions use erudite, symbolic languages which are almost incomprehensible to anyone not brought up in the relevant traditions, and so both suffer from being grossly oversimplified in popular conception: religion is generally conceived theistically in terms of worship to a creator God; science is generally conceived nihilistically in terms of perfectly mechanical determinism. When distilled for mass consumption, the most subtle, most beautiful, most compelling messages of both systems have become prostituted in favour of easily digested clichés. The profundity of religion is hidden from science. The profundity of science is hidden from religion. It is hardly surprising then that the relationship between science and religion is commonly held to be an antagonistic one. We have been indoctrinated to believe that their world-views collide, but the truth is really very different. Their maps, as we shall discover later, are not necessarily contradictory. They can support a shared mythology.

The great service of science has been to liberate us from religious dogma, clearing the board for the drawing of a far more authentic kind of metaphysical map than has ever before been possible. The trouble is that, in parallel with this emancipation, we have become so entangled within the substitute dogma of scientific materialism that no such map has yet had the chance to emerge. More than just denying us answers to those fundamental problems of existence to which the Church has always claimed knowledge, the harsh ideology of classical science actually denies the validity of the questions. With the old landmarks once provided by the Church having effectively been drowned, engulfed by the new scientific rationality, Western humanity has found itself without a compass to steer by. Indeed, the very detailed metaphysical map offered by science lacks perhaps the most vital landmark of all—a point labelled *You are here*. Our mythological adoption of this map has meant that we have cast ourselves off from our former moorings, to be lost, set hopelessly adrift in a vast open sea of rational knowledge.

Today, much storm-tossed, we are still drifting on the same rational currents, still lost. To relieve the boredom, to keep ourselves amused, we put all our time and energy into the acquisition of wealth. Instead of investing our meaning in God, we now invest it in property, and not just in property as 'things'; the possession of property extends to status and territory and people, and therefore to power. These are the idols that claim the devotion of contemporary society. The world-denial of traditional Christianity has been exposed as a con-trick. The masses can no longer be oppressed using the sedative promise of eternal salvation as reward for earthly surrender and servitude. With the 'spiritual' discredited, people are now seeking material gratification with a vengeance. First alienated from God, then from nature, Western humanity, in response to its perceived loneliness, has inevitably retreated into the ego to hide from reality behind a barricade of personal possession. Consumerism has become today's boom religion. And it *is* booming. It is winning converts all over the world, and at an unprecedented rate. It is the most contagious religion our planet has ever seen. Unfortunately, as a religion it is both discriminatory and unauthentic. Membership is not open to all, and never open on equal terms for all. Circumstances of birth and upbringing conspire to ban a great many from joining as full members. And the meaning it provides is counterfeit, an empty shell which cannot satisfy our spiritual appetite.

Our innate need for a frame of reference and a direction in life, our need for values, represents a deeply rooted spiritual desire, a desire for an authentic sense of meaning which cannot be denied without considerable trauma. We are currently living through a painful period of adjustment—our spiritual adolescence—in which humanity is trying to complete the transition from an objective ground of meaning to a subjective ground. It is proving tricky because of the almost intractable problem of communication at the subjective level. The rapid pace of our technological development can be understood in terms of the ease with which we've been able to communicate facts; the languid pace of our psychological-cum-spiritual development can largely be understood in terms of the very great difficulty we have in communicating value. There is a lot of catching up to do.

Such a huge disparity between our technological and spiritual maturity has made for a very sick society. Indeed, our loss of external value has become pathological. The malignant spread of drug addiction and senseless vandalism certainly stands as the most obvious symptom, particularly among our most deprived communities, but the crisis is really pandemic. There are more pervasive, more subjective symptoms which appear right across the social spectrum, revealing that, even for the privileged, today's metaphysical maps are woefully inadequate. As the popular side of our cultural life has drifted toward emptiness, chronic boredom and depression have become rife. We are mired in mediocrity. Indeed, a modern-day plague has infected the peoples of the industrialized world. An insidious sickness of the human spirit. An emptiness of the soul. Meaninglessness.

Our political systems make no attempt to address this root problem for the simple reason that they disown it. Acknowledging only our outward needs, they seek to create a wealthier and supposedly better society through the continued pursuit of more and more technological growth and industrial production. The mythology of industrialism dictates a policy of unrestrained economic progress, the naive belief being that a fully affluent society, with its every material need satisfied, will be free from frustration and dissatisfaction. This is, of course, an absurd reduction of the problem of individual human well-being. Material wealth is no cure for spiritual immolation. It may act as a palliative, but it is not a solution. It cannot hold back the tide of meaninglessness which is seeping through at all levels of society. Increasingly disillusioned with incompetent and irresponsible leaders who exude nothing but empty rhetoric and complacency, more and more people are becoming aware that 'something' is missing in today's culture. There is an intensely felt need to connect with some sense of the sacred. It is an innate and irresistible need. It cannot be denied.

The materialist vision is a fantasy. As the emphasis on possession of property increases, so does isolation and alienation. The more our societies become addicted to having, the more divided and fragmented they will become. Today's unashamed consumerism is disintegrating our common human spirit, and will continue to do so until we can capture some unifying vision, some common sense of meaning that will pull people together. It is a tragic irony that throughout history it has invariably taken a war to bring about such circumstances. The obvious example that comes to my mind is the spirit of the British people during the Second World War. I was not alive then, but I know well that many look back on those desperately hard times of suffering and sacrifice with astonishingly fond memories. I don't find it difficult to understand why. Humanity has always shown a quite remarkable capacity to endure suffering when driven by a greater purpose. With an authentic sense of meaning, people are able to find greatness in themselves; without it, there is often little but pettiness and weakness.

The religion of consumerism is a counterfeit religion because it only distracts us from a confrontation with our existential question. Its creed is escapism. It seeks to divert our attention from our spiritual impoverishment. It can do nothing to alleviate our deep hunger for an authentic sense of meaning. During the last war the British people did have such a sense of meaning and it provided them with a great sense of dignity, a quality which is largely missing from contemporary society. For how can we lay any claim to dignity when we are desecrating the planet on which we live, when we continue to sit idly by and let our poisonous industries defecate into our rivers and seas, and into the atmosphere—that fragile veil of life-supporting gases upon which all living systems depend for their nourishment and protection? We can surely only feel shame. The religion of consumerism is a counterfeit religion because it is destroying our original ecological intimacy, the reverence with which we once regarded the Earth, and which once provided us with a vital sense of place. This surrogate religion has led us to forget our belonging to the Earth.

However, as we enter the final decade of the millennium, a remarkable transformation in awareness has started to blossom. The signs of environmental stress, spotlighted throughout the 1980s by organizations like Greenpeace and Friends of the Earth, are finally being openly acknowledged. Among people today, there truly is concern. There truly is some willingness to begin paying the *real* price of our luxurious way of living. There truly is a desire to start making compromises. We are gradually remembering, gradually recalling our sacred belonging. But there is a very long way to go. Unfortunately, simply being aware of our problems does not, by itself, do anything to help solve them. Our concern is not enough. The currently fashionable 'greening' of the political agenda is mostly only skin-deep, motivated far more by self-interest than by a genuine ecological conscience. Swamped by a deluge of media attention, there is a grave danger that this initial wave of enthusiasm will not have the strength to break over the new barriers of complacency that are likely to be very quickly erected in place of the old. There needs to be a deeper commitment.

Only by recovering our reverence for the Earth will we be able to make any kind of genuine progress towards solving our global environmental problems. This will demand a spiritual commitment, a commitment of great courage, and great energy, for almost every bureaucratic structure within Western society is geared into the runaway drive wheels of the great capitalist machine. With all our major political parties blindly faithful to the cult mythology of Progress, with state education being shut in a secular closet, unable to address our children's spiritual imagination, unwilling to embrace the sacred in any realistic, honest way, today's commercial missionaries are too easily able to win converts to the profane religion of consumerism. It is all just too attractive for those youngsters starting out in life, seeking membership, and far too good to give up for those who are already paid-up members.

Of course, mostly, we don't have any real choice in the matter. It is virtually impossible to live in today's brutally competitive world without in some way partaking of the materialist sacrament. And it is addictive. Because the religion of consumerism is unable to offer any authentic, long-term satisfaction, we inevitably get hooked on its short-term thrills. We continually find ourselves being teased by the advertisers as they try to manipulate our perception with their glossy images and slick commercials, resolving our wants into needs, transforming luxuries into necessities. We find that we cannot help ourselves. We are tacitly coerced into joining the mad and often unscrupulous rush to consume. But it is all so very undignified, and all so futile. We *must* be capable of more than this. If we are to survive, we are going to *have* to be capable of more than this. As the cancer of industrialism spreads around the world, the Earth is beginning to sicken. And as the Earth grows sick, so do we. Physically sick. Psychologically sick. Spiritually sick. Of course, in time, geological time, our planet will surely heal itself. But we haven't got geological time to play with. Nor have our children. For their sake, we have urgently to initiate the healing process ourselves. We dare not allow the Earth to sicken any further.

Intellectually at least, we should all understand this now. The media industry has certainly done enough to plant the knowledge firmly in our minds. But, probably because it is simply too vast, the reality of it has not sunk in. It hasn't yet touched our hearts. We don't really *feel* the threat to human life on Earth because we are habituated to see only with the eyes in our head, to see just the material surface of things. Our ecological crisis is truly a crisis of vision. A crisis of perception. And it is the nature of this more fundamental crisis that I want to explore in this book. The realization of my *Earthdream* is very much dependent upon our ability to learn how to open and see with the eyes in our heart, to see beyond the superficial, to the deeper reality of things. For only when our values are conditioned by just this kind of intuitive, spiritual perception will we be able to summon the will to start living with the Earth, in a way that sustains and celebrates the integrity and diversity of life, rather than destroying and dese-crating it—which, tragically, is exactly what our industrial culture is doing at the moment. Genuine cultural transformation can only be predicated upon a transformation of our fundamental belief system. A new spirituality and morality can only be built upon a whole new rationality, a whole new way of perceiving our relations within the world, a whole new way of understanding the meaning of reality. They rise up together, a new mythology.

In the Biblical tradition, the rainbow has always been given a very special significance as the symbol of the covenant made by God with all the living creatures of the Earth after the great flood, a token of God's promise that they would for evermore be protected against such a terrible devastation. But we must now understand that we have a major role in this covenant. We, humanity, have to accept responsibility, at both the individual, personal level and at the social, political level. There is only one Earth. We have only one home. It urgently needs our care. As one global community, as one truly united society of nations, the time has come for a new covenant to be made. We have to give the whole living Earth its own Bill of Rights, to ensure, as best we can, that it will be protected for evermore, not so much from the abuse of human individuals, but from the far more powerful, inhuman, autonomous bureaucracies that they spawn.

No part of the Earth's environment, no specific portion of the biosphere, can be isolated and said to be owned by any one corporation or country or religion. The natural world is a single system. The covenant we need is a spiritual bond on a global scale, a bond that recognizes the sanctity of the Earth and is binding on all its societies of people, a bond that places the long-term health of the planet before the short-term material benefit of just one community in just one period of time, a bond declared by one visionary generation to secure the heritage of the natural world for the enjoyment and spiritual well-being of the countless generations that will hopefully follow.

I suggest that a green rainbow would be a rather appropriate symbol for such a bond. In fact, it would serve as more than just a token of this covenant; it would also function as a metaphor for the integrated, sympathetic, sustainable relationship that, very soon, we will be compelled to seek with the Earth, the perfect ecological communion which, although fully understood to be an ultimately unattainable goal, we must nevertheless always be striving to reach—a rainbow destined ever to be chased, and therefore destined ever to provide us with a source of profound meaning. The rainbow proves to be a particularly appropriate symbol because, apart from its special significance in the Judæo-Christian tradition, the idea of the rainbow as some kind of supernatural bridge between the world of the gods and the world of humanity has been part of mythologies throughout the world. It is just this same concept of a bridge between the spiritual and material dimensions of human experience that needs to be incorporated into any Green mythology.

But there can be no simple prescription for such an intangible connection. Each person's rainbow is unique, a personal way of relating the inner and outer worlds of life, a personal ethic, a personal sense of the sacred, a personal response to our existential insecurity—a rainbow for experiencing rather than possessing, a rainbow for chasing rather than catching. There is no crock of gold to be found at the journey's end. The only reward is the journey itself. I consider our individual striving toward a more loving, compassionate and moral existence and our socio-political striving toward a more harmonious, sustainable way of living in nature to be complementary and mutually supporting. I want to suggest that they represent the subjective and objective poles, the interior and exterior aspects of the same green rainbow.

The Earth is the wellspring of all that constitutes our humanity. It provides the nurturing context of both our entire physical being and our entire spiritual being. It is the sacred centre of our whole existence. It *is* our existence. To attempt to stand apart from the natural world, to attempt to divorce ourselves from this reality, is to cut ourselves off from our own being. This is the spiritual withering felt by the alienated consciousness, the homelessness, the sickness of the Western soul. This is the root cause of the West's addiction to boredom, the reason for its crisis in meaning. This is why we are imperilling our own existence through an insane disregard of the environment that sustains us. We have cut ourselves off from our own roots.

I want to see the Western consciousness recover its ancient roots in mother Earth. I want to see it recover its spiritual source. I want to see the West leading the way into a new mythopoeic era, an era which sees the creation of a mythology for the whole of humanity, for the whole Earth, to replace today's pathogenic and redundant myth of self-perpetuating material progress. I want to see this because I feel my moral obligations are not just limited to the present. They extend into future time, toward future generations. I feel a moral obligation to my future kin to leave as legacy a world just as rich in wonder and beauty, and in potentiality and diversity, as the world into which I was born. I don't much like the idea that my grandchildren might come to look upon me as a representative of the generation which blighted and desecrated the planet for the sake of one vast binge of meaningless material consumption. I would like to hope, rather, that they might come to view my generation as the one which secured the future of the Earth for *their* grandchildren. Quite simply, I want nothing less than for my generation to be the visionary generation which declares this covenant between humanity and the living Earth, this Bill of Rights on behalf of the planet. *This* is my Earthdream.

* * * * *

In order to understand what I mean by our crisis of perception it is important that we first understand something of the nature of the pattern that we unconsciously impose upon our perception of the world. And that will involve us in an understanding of the two principal forces which have shaped that pattern—our religious and scientific traditions. It is with the scientific tradition that we shall start. I want to argue that the Western consciousness is presently to be found in a kind of metaphysical stupor, still recovering from the concussion it received during its epochal confrontation with the nihilism implicit in the classical scientific rationality of Descartes and Newton. Amidst the epistemological chaos created by that mind-jolting encounter, its perspective on reality has remained obscured by a fundamental flaw at the very heart of its belief system, a flaw that is invisible to the intellect and which has consequently gone virtually unnoticed throughout the entire span of the scientific age. I am referring to the dualistic order that we habitually impose upon reality, the naive construction which organizes the universe into subject—me, *in here*—and separate object—the external world, *out there*. Functioning as an almost inviolate axiom underpinning the whole structure of Western life, this belief has come to completely shape the way in which we perceive the world and hence the way in which we also relate to and live in the world. Subject has stood against object, observer against observed, mind-stuff and material-stuff in opposition to each other, held to belong to two distinct, independently existing realities. It is in this absolute division that we have created between the inner life of humanity and the outer life of nature that we must look to find the root cause of our feelings of isolation and alienation from mother Earth.

The aim in the first section of this book, then, is to try, as far as is possible with mere words, to question and challenge the legitimacy of this division and its role in patterning the Western

25

perception of the world. We will be required to interrogate some of our most deeply engrained and cherished preconceptions of reality, and, to this end, it will be necessary to journey a little way through the surreal landscape of modern physics, the enigmatic magic of which can only be properly appreciated by pulling on our stoutest mental hiking boots and getting stuck into the trail. I can warn you now that it won't be an easy stroll. For that reason, I hope you will set off with the outlook of a genuine traveller, prepared to get involved, ready to take ideas on board and actively bounce them around rather than being merely content to look on passively from a distance. The traveller's sense of adventure and participation brings rewards that lie beyond the reach of the casual tourist. And perhaps it is just this sense of participation that I most want to capture—in a much wider context. There are far too many tourists on spaceship Earth, and not nearly enough travellers. There are just too few people who feel that they really belong on this sacred little planet of ours. And that, quite simply, is why our future is threatened. Instead of taking joy from being in the world, and therefore wanting to experience and take care of it, the majority of people in the West, particularly those people of power and privilege, expend all their energy on having the things of the world, and consequently tour through life, uninvolved, irresponsible, unaware, intent just on collecting souvenirs. Unfortunately, there is a limit to how many such fare-dodging passengers the Earth can carry.

I

The Mythology of Science

S UCH IS THE VEIL of ignorance and confusion which surrounds
the meaning of science that it would be no exaggeration to say
that popular understanding owes a far greater debt to the eighteenth
century than to the twentieth. Science has fallen prey to the modern
cult of trivialization. Its most subtle and profound concepts have
become blurred out of context through the distorting filter of
simplification. In this first chapter, then, I want to explore a
tentative path out of the confusion by trying to cut through some
of the preconceptions which have come to obscure its meaning,
both inside and outside the scientific community.

It is important that we begin by making a distinction between
pure science and applied science. Pure science is understood here
to be our systematic, empirical inquiry into the fundamental
principles which govern the organization of reality. Applied science
is basically that vast programme of research which uses these
principles in order to solve practical problems. It is pure science
with which this chapter is most essentially concerned because this
is the science which shows applied science how it should look at the
universe. Pure science provides applied science with its rationality,
the ever more effective and ingenious application of which has made
possible the creation of today's incredible technological world. And
indeed, it is in our incredible technology that we see mirrored the
power of the rationality of pure science. However, pure science is
not to be confused with technology nor indeed the use of
technology. Pure science simply conceives the possibilities of
technology. How those possibilities are actualized is down to
applied scientists, and how the resulting technology is used, or
misused, is very much down to *us*.

Founded nobly on the principles of scepticism, critical analysis and an indomitable faith in the power of reason, the story of pure science, written by a vast team of brilliant and visionary thinkers, has been one of virtually unqualified success. Unfortunately, though, largely as a result of that very success, it has also been a story soured by a considerable degree of arrogance. There has been a tendency for pure scientists to get a little carried away with themselves and equate their powerful descriptions of reality with truth. Indeed, worse still, there are many who advocate that they are to be equated with the 'whole truth' and 'nothing but the truth'. It is on this basis that science provides the rationality around which modern Western society is organized. And it is the validity of this basis which I most essentially want to question in this chapter. Let me begin by quoting the physicist and philosopher Fritjof Capra, from *The Turning Point*.

> Our culture takes pride in being scientific; our time is referred to as the Scientific Age. It is dominated by rational thought, and scientific knowledge is often considered the only acceptable kind of knowledge. That there can be intuitive knowledge, or awareness, which is just as valid and reliable, is generally not recognized. This attitude, known as scientism, is widespread, pervading our educational system and all other social and political institutions. When President Lyndon Johnson needed advice about warfare in Vietnam, his administration turned to theoretical physicists—not because they were specialists in the methods of electronic warfare, but because they were considered the high priests of science, guardians of supreme knowledge. We can now say, with hindsight, that Johnson might have been much better served had he sought his advice from some of the poets. But that, of course, was—and still is—unthinkable.[1]

The Scientific Age was effectively launched by Isaac Newton when he formulated the practical ground rules of the great game of science. Before that time there existed two distinct epistemological traditions: first, the empiricist tradition, with its origin in Plato and its Renaissance development in Francis Bacon, holding that knowledge is to be gained inductively through systematic generalizations derived from interrogation of the external world; and secondly, the rationalist tradition, with its origin in Aristotle and its Renaissance development in Descartes, holding that knowledge is to be gained deductively through human reason working from first principles. Newton cemented together these twin epistemologies into the experimental methodology of science: interfere with nature in some way; observe, measure and collect data; ruminate and reason a while; formulate an explanatory model or hypothesis and use it to make a prediction; finally, test the prediction and the validity of the model by further interference and observation. After Newton, science became defined as experimental philosophy, the practical investigation of the organization of reality. The two sides of the Cartesian dichotomy were assumed to obey the same universal logic of reason. The mind, sitting alone in the world of quality, using the tool of reason, was granted the power to discover the laws which organized the separate, external world of quantity, laws which could be described using the language of reason—mathematics.

We can see this perfectly illustrated in the example of gravity. Newton's 'discovery' of the law of gravity was just the discovery of a law and only that—the discovery of a rule of organization. The mathematics represents nothing more than a quantitative description. And this has since become the principal goal of pure science: to describe. Following Newton, the why of gravity, and even the how, became irrelevant. Its effects could be observed and measured. Its law could be formulated, tested, and subsequently verified. It simply did not matter that it was nowhere to be found, that there was no feasible mechanism. The predictive power of the mathematics was all that was important.

And, as far as physicists are concerned, this is still the case today. Developments in mathematics have provided the reasoning mind with an ever greater capacity for generalization and synthesis, enabling ever more remarkable feats of description to be performed. Through its gradual working out of the mathematical rules by which the physical world is organized, pure science claims to be modelling and explaining the universe—totally and completely.

Indeed, the unashamedly arrogant goal of today's front-line physicist is the unification of these rules into one simple mathematical formula. In this ultimate equation would be found the explanation of the universe. All that is, the whole of reality, supposedly, would potentially exist within the internal mathematical logic of this one rule. But, as we sit today on the threshold of this tremendously exciting and momentous landmark of human intellectual endeavour, perhaps we should take a few steps back to obtain a broader view. Does this mathematical description really represent an explanation of the whole of reality? And where exactly is the world of quality, the world of consciousness, to be found in this magical equation?

At the end of the seventeenth century, classical science stood like a steamroller at the top of a very long hill. Newton released the handbrake. As its inexorable internal logic worked itself out through the great scientists of the eighteenth and nineteenth centuries, the steamroller squeezed the world of quality right out of the picture. Descartes' dualism was flattened into a monism. The way was made clear for the mind to be relegated to the status of an incidental feature of the universe, no longer a separate world at all. It came to be regarded as just an epiphenomenon, a curious side-effect which had arisen merely as a consequence of the remarkable complexity of the world of matter. This fairly accurately represents the prevailing view of mainstream science as it stands today.

But there is a bewildering paradox inherent in this view. Science seems somehow to be sitting under the cast of its own shadow. The model produced by its rationality has no place for the source of that rationality—the reasoning human mind. Our very concrete world of consciousness, the living world of thought and sensory experience, the world into which we are born, the world from which we die, has been reduced right down to the bare geometry of nothingness, a substrate of abstract mathematical law. There is absolutely no logical necessity for the existence of consciousness in the symbolic model of the scientist. Nothing would change if the light of our consciousness was suddenly snuffed out: the entire universe, according to science, would carry on its business exactly as usual. In the words of Erwin Schrödinger, one of the great pioneering physicists of this century:

> On the one hand mind is the artist who has produced the whole; in the accomplished work, however, it is but an insignificant accessory that might be absent without detracting from the total effect.[2]

Although science tries to distance itself completely from any kind of metaphysical speculation, it still cannot avoid the fact that its theory, at the most fundamental level, is actually embedded in metaphysics. Certain metaphysical assumptions have necessarily to be made before science can proceed in the first place. One of these, for example, is the assumption that local rules can be applied globally, the assumption that what is discovered in the laboratory, at atomic scales, at one particular moment in time, can be applied to the whole universe, across all scales, and across all epochs. There can be no proof that this assumption is valid. Like all such beliefs, it simply has to be laid down as a metaphysical postulate.

The two most important of these *a priori* postulates are the closely related Principles of Objectivity and Reductionism—both a direct legacy of Cartesian dualism. The Principle of Objectivity is the axiomatic belief that the external world *is* external, that it can be reduced to an object of study existing independently of the human mind. It assumes an objective reality of separate, countable objects which can be quantified and described by a mysteriously detached human consciousness. The Principle of Reductionism is the axiomatic belief that the whole is identically equal to the sum of its atomistic parts, that every aspect of reality can be reduced exclusively to the soulless logic of the mathematical laws which govern those parts. It assumes that the explanation for every physical and mental phenomenon is ultimately to be sought at the atomic scale, in the interactions of the most fundamental particles and fields of physics. Here, then, in this pair of metaphysical beliefs, we reach the generating instrument of science, its very cutting edge, its rationality. We find the clinical scalpel of reason, with surgical precision, splitting the world neatly into two, pairing reality into objective and subjective. This is the value which creates and validates what we normally understand as scientific knowledge. It is also the value which invalidates all other kinds of knowledge, most significantly intuitive knowledge.

Instead of just being used to form the most effective working context of the scientist, an essential framework of convenience, the Principles of Objectivity and Reductionism have been elevated into an ideology, the one absolute instrument for the investigation of reality, the one rationality to be used by all scientists. Thought, emotion, sensation, intuition, all are just physico-chemical activities of the brain and body, just matter-energy processes. The rationality of science dictates that only the physicist, using mathematics, can offer an insight into the organizational principles of the universe. Physics is therefore seen to be the only pure science. The study of human consciousness, four orders removed from centre stage, can offer no such insight; psychology is just applied biology, which is just applied chemistry, which is just applied physics, which is, essentially, just applied mathematics. The hierarchy is held to have one incontrovertible direction of epistemological flow. But again, there can be no proof that this is the correct way of looking at the order of reality; it is simply part of the belief system of what I shall in future refer to as traditional science. And, while acknowledging the amazing success of this system as a tool for understanding the universe, it is important that we keep very much in mind the fact that it is *only* a belief system, and that it has a number of inbuilt limitations.

The most obvious of these limitations arises from the fact that wholes can never be analysed as wholes: biologists have to dissect living organisms; chemists have to disassemble molecules; physicists have to split atoms. And by taking something apart one is, of course, immediately precluded from asking certain types of question—perhaps the most important type. By way of analogy, let us consider a very familiar example. The rationality of traditional science allows it to investigate the function of a television set in intricate detail; it can take it apart and precisely analyse and describe the workings of all its bits and pieces—but it cannot analyse the television *picture* since it is nowhere to be found among the scattered array of mechanical parts. The picture can only be obtained through the purposeful functioning of the whole set, just as our experience of the colour green, say, can only be obtained through the purposeful functioning of our whole brain. The fully functioning television set is necessary to ask the question to which the television picture is the answer. Because the objective, reductionist methodology of traditional science cannot provide the tools required to fashion such subtle questions, it is clearly prevented from embracing a certain domain of possible answers. For one, it cannot observe and describe the colour green because its rationality does not allow it to ask the right kind of question.

Scientific observation inevitably involves isolating a small part of nature by inhibiting as far as possible its interaction with any other part. It follows then that scientists are forced to create situations which are artificially restricted, and can therefore only expect restricted answers back from nature in reply to their experimental questions. Indeed, by the very design of their experiments, they must impose limiting conditions on nature, conditions which may force her to give the 'right' answers, answers that in part reflect the view of the world that they have already presupposed. One of the most obvious such limiting conditions is that the systematic methodology of science precludes the possibility of embracing rare or spontaneous events; data has to be statistically repeatable in order to be recognized as significant. This feature of scientific inquiry makes it a particularly unsuitable tool for the investigation of random subjective phenomena involving the role of human consciousness. Once again, there exists a whole domain of possible answers to which traditional science is denied access because its rationality lacks sufficient subtlety. This is not a criticism, simply a limitation that must be acknowledged. In the words of Werner Heisenberg, another of the great pioneering physicists of this century:

> We have to remember that what we observe is not nature in itself but nature exposed to our method of questioning.[3]

There are just three essential criteria that need to be satisfied in order for a method of questioning to be labelled as scientific: first, an unprejudiced, empirical, systematic investigation which yields a well-defined set of data; secondly, the creation of a self-consistent model which attempts to explain the data; and thirdly, and most importantly, a further rigorous and systematic investigation which is directed towards trying to falsify the proposed model. These conditions certainly limit the scope of the scientific method of questioning, but they do not dictate that the process has solely to address quantitative measurement to the exclusion of qualitative experience. They do not deny the validity of a science of shared, subjective experience. Psychology, because it can use the miraculous tool represented by human consciousness, can ask questions which cannot be framed using the instrument of analytical reason, quite possibly giving it authentic insight into aspects of reality from which traditional science has cut itself off, aspects perhaps characteristic of some very much deeper and more profound organizing principle. Science should properly include all possible means of insight within the compass of its inquiry—an inquiry which should also be turned inwards to question its own values, most specifically its own fundamental metaphysical assumptions.

In a way, scientific observation can be likened to looking for pictures in a fire, as opposed to, say, the passive operation of looking at a photograph. For scientists play an active role, a role in which it is *they* who create the image. What scientific facts are picked out of the flickering chaos depends very much upon what is already believed to be there. All scientific theories are ultimately underpinned by value judgements. And the construction of scientific reality from these theories is agreed by common consensus using the same principles. The largely unacknowledged problem is that the consensus for the traditional construction of scientific reality is agreed using the value of reason, premised upon preconceived notions of what actually constitutes that reality. The images that are picked out of the fire are self-selected by virtue of the fact that traditional science sees through the eye of just one rationality. It is as if the rules of the scientific game have been fixed in advance to guarantee the triumph of this one particular way of looking at the world.

The ideology of science is self-validating in much the same way as traditional religion has always been. As Christianity has the Bible,

as Islam has the Koran, so science has the Principle of Objectivity—
the outrageous implicit assumption that underpins the traditional
scientific method, unexposed, unexamined, unquestioned. But it is
not a truth; it is a belief, a belief which we are supposed to accept
as an article of faith, and, for many, as an article of blind faith.
Indeed, it is only a very few people who really understand what it
is that they are actually being asked to believe in. Most of those who
have adopted the map of scientific materialism, not having the
intellectual determination to grapple with the epistemological
problems of science themselves, simply trust to the validity of the
scientific creed. Certainly, there is a necessary place for faith in this
strange universe of ours, but not a blind faith. In this respect, there
really is no difference between the faith of science and the faith of
orthodox religion, except that scientists would argue that their faith
is rational while that of the Christian or Moslem is irrational. But
of course, by such an argument they are simply using their own
sacred rationality to justify their faith in that rationality, just as
many Christians use the Bible to justify their faith in God.

This faith is administered by the great academic establishment
of science, the voice of a committee of opinion backed up by all the
weight of tradition and respectability and prestige. Although at a
personal level individual scientists are often free-thinking and
open-minded, with many holding unconventional beliefs; at the
collective level, the faceless, bureaucratic establishment of science,
anchored by the inertia of its ideology, assumes an iron mask of
unimpeachable objectivity. It conceals the subjective aspect of its
dialectic for fear of undermining its venerated status as the guardian
keeper of supreme knowledge. And this is, indeed, the position it
has come to occupy. Science has assumed the authority that was
once vested in the traditional Church, and in the process has also
taken aboard much of its imperialism, its institutions acting as the
arbiters in deciding which scientific theories to recognize and
uphold. The accepted theories become absorbed into the institutionalized view of the world and are subsequently protected by
forming part of the sacrament delivered to the next generation of
scientists in schools, colleges and universities. The new generation
inherits this mythology as the received dogma, with very little

freedom for questioning and criticism; the ranks soon close in on any student not content to toe the establishment line. Tradition and conformity exert a totalitarian rule over radicalism, for academic authority cannot afford to foster an approach in its students which might lead to doubt being cast on the validity of its own creed. Nothing must be allowed to devalue the status of scientists as our high priests of knowledge.

This scenario may have been a little exaggerated for rhetorical effect, but probably not as much as one might think. Although the vehicle of scientific progress is driven by ideal, it is still largely steered by tradition. Today's scientists inherit a large body of laws and theories, the legacy of many thousands of elegant and ingenious experiments carried out over a period of some 300 years. These form the context, the tradition, in which they make their observations. The observational data from their experiments is then interpreted within the framework offered by this context. Instead of keeping themselves open to the widest range of possibilities and inputs, scientists inevitably proceed with conservative, dogmatic blinkers. Science is not performed in a vacuum. The development of science is inextricably embedded in the historical process and it cannot be abstracted and analysed out of that context. Its principal movements are now seen to be reflections of movements of thought within the general matrix of society. And there is no reason to think that this relationship has changed in any way, especially now that scientific research is so expensive, and the funding must come from government. Only the 'right' kind of science is financed. Only the 'right' sort of experiments attract research grants. The incestuous relationship of science and the State ensures that only the 'right' questions are asked. Their institutions have become mutually legitimating. Their ideologies have become mutually reinforcing.

Certainly, on the face of things, traditional science seems to be very well justified in wielding its analytical knife with such great confidence and certitude. It represents a highly successful idolatry. It has swept all before it. It has placed unprecedented power in the hands of the State, which has in turn given tremendous power back to science in terms of status and prestige. It has become an awesome partnership. But its effectiveness has largely been due to the fact that it only addresses the domain of quantity. It is not the truth or exclusivity of its knowledge that validates traditional science, nor indeed the cogency of its methodology, but simply the power that its knowledge wields through technology. Pure science has no power of its own; it is only seen to be powerful because its rational knowledge has put so much power into the hands of technologists. The validity of science, as measured in this way, cannot and must not be denied. Nature cannot give false answers to the experimental questions of the scientist. However, what needs to be freely admitted—and what my polemics have basically been all about— is that traditional science, by virtue of its own ground rules, can only provide us with a partial view of reality. It must inevitably close its doors to alternative forms of knowledge and principles of organiza- tion which, because of its metaphysical self-definition, it cannot hope to embrace or even address. Unfortunately, though, the sacrosanct dogma of traditional science insists that no principle exists which cannot be tamed by the generalizing laws of mathe- matics. And this idolatrous worship of the Principle of Objectivity has allowed the dialectic between the objective and the subjective to be stuck in an ideological groove. The fundamental tenets of traditional science urgently need to be deconsecrated. It has to be acknowledged that pure scientists are not priests but poets, not uncovering truth but finding metaphors, not seeking explanations but creating a mythology.

There is no single, absolute scientific rationality, and therefore no single, absolute scientific reality. It all depends on the values we choose to build into the foundations of our model. There is no single construction that can be called correct. Any such model, created as it is by looking at the universe from the perspective of one particular rationality, represents just a partial picture. The universe is too rich in structure and content to be described on one canvas. We need two complementary pictures, created by looking at the universe from two different perspectives: the objective point of view of its parts and the far more subtle, subjective point of view of its wholes. Science, in its purest form, is the natural philoso- pher's original imaginative quest to embrace metaphorically the fundamental realities contained within the vast landscape of the universe. It is all about creating metaphors and shaping them into narratives to frame the patterns of nature as revealed through the sceptical and highly disciplined investigation of both the objective and subjective dimensions of reality.

Directed by the one singularly powerful rationality of reason, traditional science construes reality in a way which has enabled us to advance our understanding of the universe at an astounding rate. Indeed, the story of modern science is quite genuinely astonishing, and seems close to reaching some kind of astonishing climax. As a mathematician by training, I can appreciate with more wonder than most perhaps the extraordinary elegance of the scientific model of physical reality as it currently stands. In addition to having enormous descriptive and predictive power, it is aesthetically beautiful in a way that, unfortunately, cannot easily be conveyed to anyone not conversant with the language of mathematics. The prospect of these laws being unified under one principle—the so-called superforce—is therefore tremendously exciting. But although there will no doubt be many extravagant claims to the contrary, such an achievement, however extraordinary in an intellectual sense, will *not* mean that the universe has been 'explained'. It will not signal the end of pure science, only our arrival at a new pinnacle of description of just one particular, fundamental construction of reality—one which does absolutely nothing to advance our understanding of the fundamental reality of consciousness.

The classical vision of some final edifice of complete scientific knowledge is a dream castle. It is an old myth which has been rewritten in the twentieth century, and actually rewritten from within science itself. Modern physics, as we shall shortly see, has delivered some severe blows to the foundations of the ivory tower from which science has for so long looked upon the world. It has, in fact, done much to free us from the straitjacket of Cartesianism, establishing an authentic dialectic between the objective and the subjective. It has shown us that we cannot brush the phenomenon of mind quietly aside and sweep it under a material carpet. It has shown us that the worlds of quantity and quality are not to be cleaved apart as simply as had always been conveniently assumed.

The bottom line is that they are not to be cleaved apart at all. They are one world. The story that science is trying to tell has been shown to be inenarrable. It is one of the great tragedies of the century that this profound message has taken so long to impinge even slightly upon the Western consciousness. This serves to introduce a very important point with regard to the cultural role of science. Here is Schrödinger again:

> There is a tendency to forget that all science is bound up with human culture in general, and that scientific findings, even those which at the moment appear the most advanced and esoteric and difficult to grasp, are meaningless outside their cultural context.[4]

There is a trend today which confines scientists to ever tighter and narrower fields of research. There is usually little opportunity to explore other fields, principally because of the enormous pressure to produce and publish results, first, in order to appease those institutions funding their research, and secondly, in order to work their way up the pecking order within their own particular highly specialized scientific discipline. The great majority of scientists work and publish within very small, very élite coteries: physicists rarely talk to biologists; psychologists rarely talk to computer scientists. Knowledge has become increasingly ghetto-ized—at the expense of wisdom. In this way science this century has slowly lost contact with its cultural roots. I would like to argue that the scientific pursuit cannot properly be divorced from popular culture. Publishing an erudite paper in an esoteric language that is only intelligible to a handful of fellow experts is a very sterile and narrow form of science. As a rather special kind of mythic medium, science should communicate meaning, and to the widest possible audience. It has a vital cultural responsibility, through education, to make the consensus of its reality as large and as informed as possible. Unfortunately, young people at school are all too often indoctrinated with science rather than educated. This was certainly true in my own case. I was never given an opportunity to think about and question the validity of the metaphysical underpinnings of the science which I was taught. The *ideology* of science was imposed upon me, just as it was upon countless of my contemporaries, with the result that a whole generation has grown up intellectually corrupted, with a very distorted perception of the world, and of the very concept of reality.

The main aim of this chapter has been to distance the *ideology* of science from the *art* of science. If it seems to have hit home rather hard, that is simply because it is important to rescue the meaning of science from the clichés and over-simplifications that have sadly invaded our everyday understanding. Nevertheless, there is a fine balance to be sought here. In being highly critical of the central conceptual dogma of traditional science there is a tendency, perhaps, to lose sight of the essential importance of the very exacting values that guide its inquiry. If the meaning of science is broadened too liberally, there is a danger that prejudice and narrow-mindedness will give way only to deception and woolly-mindedness. To expand our understanding of the meaning of scientific rationality is not to open a door to irrationality. Sadly though, in reaction against the sterility of the world-view of traditional science, more and more people do appear to be succumbing to the demon of unreason. Our very sophisticated, hi-tech society is home to a remarkably widespread willingness to believe in what to me seems like all sorts of ludicrous nonsense. This immediately raises a very deep and fascinating question: amidst the confusion of today's world, with popular faith in the rationality of science dissolving away, just how is one to distinguish non-sense from sense?

We are fortunate to live in a society which has a free market policy in ideas. It is recognized as a basic human right that we each have the freedom to choose our own truth. The trouble is that when it comes to education, a completely open approach cannot be adopted for very obvious practical reasons. Young people require some quite specific frame of reference, some well-defined mythology, to function as a solid foundation for their common sense. What kind of mythology, then, are we to present to them? To ask this question is to step into an uncharted expanse of epistemological quicksand, strewn with all kinds of apparently bottomless questions concerning the nature of truth. At this early stage of our journey it will have to suffice simply to suggest that our present vague and contradictory mish-mash of traditional science and naive Christianity is grossly inadequate, and that in replacing it, we have to begin by looking to the wider vision of science which I have been trying to develop in this chapter. If scientists were to embrace their role as mythopoets, science could be presented to our children in a way that would make it irresistible. If it were properly appreciated that there are two interdependent scientific realities as opposed to just the one promoted by traditional science, far fewer people would feel alienated from scientific mythology. Young people need to be taught to respect the power of reason as well as to understand its limitations. And they need to be encouraged to develop the art of critical thinking. The mythology of science should be their guiding example, and that is why it is so important that it is seen to be founded on a humble and honest scepticism—not scepticism as it is popularly understood, as disbelief, but scepticism as the unprejudiced doubt which actually conceives the rigorous question and answer dialectic of scientific inquiry. People might then be inspired to doubt, inquire and discover for themselves, and within themselves—for the inquiry into our own subjective reality is no less scientific than the inquiry into the reality of the physical world.

The scientist cannot get outside nature to observe it in a detached way. We all exist as part of nature, as one organic, irreducible whole. This is why the language of science is inevitably metaphoric in character. Nature has no voice of her own. She is utterly, utterly silent. Scientists cannot hear anything but the distorted echoes of their own voices. Their art, ultimately, is the creation of metaphors, to be incorporated into a mythology which attempts to ever more authentically embrace and communicate the meaning of the patterns found within those returning echoes. Rather than being concerned with killing the universe with explanation, science is all about bringing it to life, about creating new ways of construing its essential paradoxicality. This is the inspiration of the authentic mythology of science.

2

The Physics of Participation

THE BASIC THESIS of this first section of the book is that today's prevailing scientific mythology provides us with a seriously flawed perception of our universe. And since our understanding of who we are as human individuals, indeed our very identity, is so intimately bound up with our understanding of what kind of world we live in, we tend to develop a sadly impoverished sense of selfhood. In the words of Morris Berman:

> Scientific consciousness is alienated consciousness: there is no ecstatic merger with nature, but rather total separation from it. Subject and object are always seen in opposition to each other. I am not my experiences, and thus not really a part of the world around me. The logical end point of this world view is a feeling of total reification: everything is an object, alien, not-me; and I am ultimately an object too, an alienated 'thing' in a world of other, equally meaningless things. This world is not of my own making; the cosmos cares nothing for me, and I do not really feel a sense of belonging to it.[1]

Our culture is now so immured within the fundamentalist view of scientific materialism that it has become virtually impossible for us to imagine how the world, and our alienated relationship to it, can be perceived in any other way. Which makes my task here, in trying to present an alternative perception—and, indeed, your task too, in trying to appreciate it—an extremely challenging one. Since our thought about the nature of the physical world is shaped so completely by the language that we use in its rational description, the limitations of that language unavoidably impose a serious constraint upon our ability to move toward a more authentic

43

perception. The basic trouble is that we are necessarily obliged to use words which are grounded in common-sense experience to describe a reality which has long since parted company with the reality of common sense. The most obvious culprit is the word 'physical' itself, as used to describe the world which is examined and 'explained' by pure science. The word 'physical' immediately invokes the ideas of substance and solidity, ideas rooted very firmly in our conscious sensory experience of the world. But what actually is the meaning of physicality in isolation from human experience?

Since time immemorial, philosophers have speculated about the nature of the 'stuff' which composes the material world. For the greater part of the last 2,000 years, the prevailing view was that of Aristotle. He proposed that everything in the universe is formed from just four basic constituents: earth, air, fire and water. It appears very naive now, but in fact it is easy to appreciate the natural appeal of such a theory against the backdrop of a mythic cosmology. Without a science to reduce the observations of everyday experience, Aristotle's theory made a great deal of sense—which is possibly a lot more than can be said for today's popular view. We are now taught the principle of atomism, a theory originated by another Greek philosopher, Democritus, in the fifth century BC. All objects are composed of an immeasurably vast collection of unimaginably small, invisible objects called atoms. We learn that the 'stuff' of the universe is just lots of microscopically small 'stuff'—which means that we learn nothing very much at all.

Within the classical view—which is the view that was presented to me at school, and most likely to you too—the assumption is made that the component atoms exist in the same form as that of our everyday macroscopic objects, extended in three dimensions, possessing the quality of discreteness, and having some kind of vaguely defined substantiveness. Indeed, the picture of the atom which most people form in their mind is inevitably of some kind of solid, spherical ball of physical material. But we now know that atoms do not represent the ultimate substance of nature. This century, science has provided us with an updated picture which sees the atom being built from three distinct, more fundamental building blocks: protons and neutrons clustered together to form

44

a central nucleus, around which orbits a number of much smaller electrons. Our curiosity should now compel us to inquire into the nature of the material which composes these sub-atomic building blocks, the so-called elementary particles. Science has more recently come up with the answer that protons and neutrons are composed of yet more fundamental objects called quarks. Now, of course, we are compelled to inquire into the nature of the material which composes these even more elementary particles. Clearly, we could go on like this forever. Following any such reductionist programme, we become hopelessly ensnared within an infinite regression.

It is impossible to visualize the ultimate stuff of nature in terms of our everyday experience. We cannot conceive of some basic building block without also conceiving of some material from which it is composed. Our questions, in such a simple form, can never be answered. They are merely transferred to ever more 'fundamental' levels. The concept of a discrete, material, solid particle is ultimately redundant. To try to picture the stuff that composes a quark is as meaningless as trying to picture the stuff that composes the colour green. To try to imagine what a quark looks like is as meaningless as trying to imagine what it smells like! At this level, physical extension, discreteness and solidity mysteriously vaporize before our mind's eye. They are really no less naive than Aristotle's earth and fire.

However, in spite of this, it is the idea of the world consisting solely of billions of discrete, tiny, solid particles that we sell to our children in science lessons at school. The atom is pictured as a miniature solar system with the protons, neutrons and 'orbiting' electrons 'painted' as coloured spheres sitting in an absolute space. This perversely archaic representation of the atom is readily and indelibly imprinted on the young mind. It certainly was on mine. When its limitations were eventually exposed, I found it immensely difficult to escape from its easy familiarity—in fact, impossibly difficult, for I am *still* struggling to break completely free.

We will shortly see that the image of a particle is a very misleading metaphor to use in relation to the structural components of the physical world. The trouble is, though, that there are very few viable alternatives. There are simply no words available that get even remotely close to conveying the right kind of meaning. In choosing an alternative to 'particle' then, I tried to find a word with the barest minimum of confusing connotations, eventually settling upon 'pattern'. By referring to the physicist's fundamental particles as fundamental patterns, there is an immediate and far more meaningful emphasis placed upon form over material. In all practical respects, it is only form which has any significance in the phenomenal universe. The fundamental particles of the universe are best thought of as multi-dimensional patterns traced in the fabric of space and time—or, more properly, spacetime.

Following Einstein's formulation of the theory of special relativity in 1905, space and time can no longer be considered independently of each other. Rather than leading separate lives, as immutable absolutes, space and time share in an essential fluidity. They are involved in each other's existence. A natural mathematical consequence of this strange relativity of space and time is the relativity of mass and energy, as immortalized in the formula $E = mc^2$. Einstein's theory of relativity, in a very abstract way, has revealed all matter to be reducible to energy. Ultimately, the 'stuff' of matter is the 'non-stuff' of energy.

Although we are constantly aware of its effects, energy is an astonishingly intangible concept. Energy cannot be held in the hand. Energy cannot be looked at. Energy cannot be visualized in any conceivable way. Our only hold upon it is a symbolic one—as a term in a mathematical equation. We have to conclude, then, that the 'stuff' of the universe has no actual 'physical' reality in any sense that we ordinarily understand. The most important point, though, is that the ultimate nature of this 'stuff' is as utterly insignificant as it is utterly ineffable; in fact as insignificant as the chemical composition of the printing ink to the meaning of the words that you are reading in this sentence. It is only the *patterns* that are traced by this 'stuff' that have any genuine reality in our universe. The aim of the modern physicist then is to discover the rules which govern the form of these traces. They seek to define the mathematical grammar which gives form to the physical text of the universe.

46

Their latest theories suggest that matter be pictured as 'locked-up' energy, energy that is somehow wound into the internal geometry of a spacetime of eleven dimensions—seven of which have been compacted to an unimaginably small scale. In this context, we should think of a fundamental pattern as a specific geometrical configuration of energy. The set of unique fundamental patterns in the universe can then be thought of as the set of fundamentally different ways that energy can dance around the multi-dimensional geometry of spacetime. Each fundamental pattern has a different mathematical 'shape', a different mathematical topology, a different pattern of excitation. It is the abstract mathematical properties of these dynamic forms which define the grammar of reality, giving rise to the atomic properties that are measured and quantified by physicists.

As an example, let us look at the collision of matter with antimatter, which, as is now well known, results in explosive mutual annihilation. This is nature's most spectacular interaction. All the energy of being—the combined mass of both matter and anti-matter—is converted into free, unbound energy as determined by Einstein's famous formula. The only essential difference between matter and antimatter is in the sense, or the handedness, of their excitational patterns. When a clockwise pattern of excitation meets an anticlockwise pattern, the two mathematical shapes cancel each other perfectly. Both forms dissolve into formlessness, their combined energy being freed in a mutual unwinding of their dynamic mathematical structures. Imagine winding a piece of string clockwise down a length of small rod. Halfway along, we reverse the sense of the twisting so that by the time we reach the end of the rod we have created an equal number of anticlockwise twists. Here is a model of matter meeting antimatter. Grab hold of each end of the string, pull hard, and *twang*, form gives way to a formless length of string.

At this point, before too clear a picture has a chance to emerge in your mind, I have to begin introducing some complications. Unfortunately, it turns out to be too simplistic just to swap our image of a discrete material particle for an image of a discrete pattern of energy. A lot of the trouble has to do with that word 'discrete'. Because the objects of our everyday world seem to be perfectly discrete and self-contained, we naturally want to picture the objects of the atomic world in the same way. But the truth is that the micro-world cannot be pictured in the same terms as the macro-world. We cannot indefinitely chase concepts like solidity and discreteness from the macroscopic into the more and more microscopic. There has to come a point where we accommodate some kind of conceptual shift. Quantum theory is the mathematical tool which has allowed science to probe beyond this point—and has revealed nothing but the unimaginable: massless, dimensionless abstractions like the one that is labelled energy.

Under quantum theory's symbolic illumination, we find that our familiar, everyday world of 'things' diffuses into a vacuous world of fuzzy shadows. Quantum theory strikes to the very heart of reality, but its thrust is found to meet no resistance. It reveals our world of substance to be but an impressive illusion conjured up by the magic of Mind. And I shall take care to point out straight away that Mind, as I capitalize it here, is not to be directly identified with the mind with which we each individually think. There is a very important distinction to be made, which we will be discussing in detail in the next chapter.

In his introduction to *The Nature of the Physical World*, first published in 1928, Sir Arthur Eddington offers this now quite famous statement of the puzzling situation in which we find ourselves:

In the world of physics we watch a shadowgraph performance of the drama of familiar life. The shadow of my elbow rests on the shadow table as the shadow ink flows over the shadow paper. It is all symbolic, and as a symbol the physicist leaves it. Then comes the alchemist Mind who transmutes the symbols. The sparsely spread nuclei of electric force become a tangible solid; their restless agitation becomes the warmth of summer; the octave of aethereal vibrations becomes a gorgeous rainbow. Nor does the alchemy stop here. In the transmuted world new significances arise which are scarcely to be traced in the world of symbols, so that it becomes a world of beauty and purpose—and, alas, suffering and evil.[2]

Before we go on to discuss the alchemy of Mind, and the human world of value and meaning, we must first try and understand a little more about the symbolic world of quantum physics. To lay some groundwork, which will help us to appreciate as fully as possible the bizarre paradoxicality of the quantum world, we must turn our attention briefly toward a study of light. Throughout history, thought has oscillated between wave theories and particle theories to explain the properties of light, and, indeed, the great Isaac Newton toyed with both before plumping for the idea that light is a stream of minute particles which he called corpuscles. Although an alternative wave theory had been developed by Christiaan Huygens at about the same time, the immense scientific stature of Newton ensured that the corpuscular theory stood intact throughout the eighteenth century.

However, in 1803, things were very much turned around as a result of a remarkably simple experiment performed by Thomas Young. By shining light at a screen containing two very narrow slits, Young was able to record a pattern of light and dark bands upon a plain white screen beyond. This was an observation which simply could not be accounted for using Newton's corpuscular theory of light. The pattern observed by Young is an interference pattern of two wave fronts, and it will be useful for us to understand exactly how it is formed.

To begin with, it is important to note that the light which is visible to our eyes actually represents just one small band in a whole spectrum of radiation which includes, among others, radio waves, microwaves and infra-red, ultra-violet and X-rays. All these different kinds of electromagnetic radiation are identical except for their wavelength—the length between the individual 'crests' or 'troughs' of the wave—or, what is essentially the same thing, their frequency—the number of oscillations per unit of time. It is also important to point out the popular error of picturing wave motion in terms of a discrete particle travelling a roller-coaster trajectory through space. Instead, what we really have to picture is the far

49

more subtle concept of a pattern being transmitted by the oscillations of a propagating medium.

The best common example of wave motion is that observed when a pebble is cast into a pond. The waves or ripples spread out from the point where the pebble strikes the water, but the body of water has no outward motion of its own. There is no horizontal component to the motion of the water, only oscillation in a vertical plane, transversely opposed to the direction in which the wave propagates. Nothing of a physical nature is to be observed travelling outward across the pond. What we witness is purely the transmission of a pattern of energy.

The most dramatic illustration of the wavelike nature of light is, of course, offered by the rainbow. Another example, and one which is slightly easier to explain, is the closely related phenomenon of iridescence, that beautiful interplay of suffusing colour to be seen in a soap bubble or a thin film of oil. Here, light is reflected from both the top and the bottom surfaces of the microscopically thin film, with the result that the two reflected waves interfere with each other. We have to note that ordinary white light is a combination of a whole range of waves, each with its own distinct wavelength. Some wavelengths will be such that the crests of the two reflected waves coincide with each other. The oscillations of the propagating medium—which in this case is the fabric of spacetime itself—add up. The waves *constructively* interfere with each other. Other wavelengths are such that the crests of one reflected wave coincide with the troughs of the other. The oscillations of the propagating medium cancel each other out. The waves *destructively* interfere with each other. In this way, the white light is split among the various parts of its component spectrum, presenting an ever-shifting image to the eye as it moves in relation to the reflecting film.

Young's interference pattern arises in much the same way. Because the width of the slits is comparable to the wavelength of the light, any wave incident upon them continues to propagate as two separate waves, centred upon the two slits, each spreading out in space beyond the plate in the same way that ripples spread out upon a surface of water. As a result, these two new wave fronts inevitably interfere with each other. Again, there will be both constructive and destructive interference. There will be places where the waves are superposed upon each other—recorded as bands of light—and places where the waves cancel each other out—recorded as bands of dark.

At the end of the nineteenth century, with Maxwell's elegant mathematical formulation of electromagnetism complete, the wave theory of light was set as firmly in place as Newton's corpuscular theory had been in the previous century. However, the unique genius of Einstein turned things around again when, in 1905, in a paper appearing alongside his paper on special relativity, he proposed a new corpuscular theory in which light is viewed as a stream of discrete packets of energy which have come to be called quanta or photons. Photons are the simplest fundamental patterns to be found in nature because they have no mass, nor any other complicating characteristics. They represent free energy unlocked from any bound configuration, a fact which explains why they are always to be found travelling at the same fantastic speed. Possessing no inertia, they are unable to resist being instantaneously accelerated to nature's limiting velocity of 186,000 miles per second—a velocity which is defined by the particular way in which space and time interrelate. Applying his theory of quantization, based on the earlier work of Max Planck, Einstein was able to solve some puzzles concerned with the interaction of atoms and electromagnetic radiation which made no sense using the wave theory.

Light could now no longer just be thought of in terms of being a wave *or* a particle. Light is both wavelike *and* particle-like. This is a very strange alliance of characteristics. Remember that, as a particle, a photon is a lump of energy which is localized at a *discrete* point in space and time, but, as a wave, it is spread *continuously* throughout space and time. Think again about the ripples on a pond. The wave is actually the *whole* motion of the water surface as it changes with time. It is as meaningless to talk about the location of the wave as it is to talk about the frequency of the pebble! But, if waves and particles are such alien principles, how is it that a photon can incorporate the characteristics of both? Not in any way that we can describe in words or pictures. However, by way of some bewilderingly abstract mathematics, this wave-particle duality can

51

be captured with stunning simplicity. Indeed, in one glorious period spanning the years 1925–6, three quite distinct mathematical descriptions of quantum behaviour were developed in turn by Werner Heisenberg, Paul Dirac and Erwin Schrödinger. It eventually turned out that Heisenberg's description concentrating on the particle aspect of quantum duality and Schrödinger's description concentrating on the wave aspect were but special cases of Dirac's complete description which integrated these dual aspects in a sublimely elegant, but at the same time diabolically obscure, way.

The trouble is that the symbols in Dirac's equations do not represent anything upon which we can place familiar labels. They have absolutely no analogies within our realm of common experience. Quite simply, they have no physical counterparts. However hard we might try, it is just not possible to capture these patterns in a mental picture. To try to visualize what a fundamental pattern looks like is to fall instantly into an epistemological trap. Vision is as inappropriate to the quantum landscape as sound or touch. We have to leave behind the idea of discrete and solid 'things' with well-defined, objective characteristics and fixed unambiguous positions in space and time. These concepts are just not valid at the quantum level of description.

To explore the paradoxicality of wave-particle duality further in all its puzzling splendour, I would like to return to Young's slits experiment, although this time we will consider electrons instead of photons to emphasize the fact that quantum duality applies equally to all the fundamental patterns of nature, even those configured with considerable mass. The experimental situation described here is a slightly idealized one in order to make the presentation a little easier, but this has no effect on the essential principles involved. This time we direct a beam of electrons at the slitted screen, behind which is placed a large array of Geiger counter-type detectors, each with its own numeric display. What do we find?

After firing our electron gun continuously for a few seconds, an analysis of the displays will reveal a series of vertical bands where very high counts have been recorded, interspersed with bands where very low counts have been recorded. The natural conclusion is that we have registered an interference pattern. The electrons appear to be behaving like waves and are interfering with each other in the way light waves do. However, they are also behaving like particles, for each electron appears to arrive quite discretely, triggering a single click at just a single Geiger counter, striking our array of detectors as if it were a bullet being fired from a gun.

We could crudely compare this result to a situation where a pebble is dropped into the very centre of a circular tank of water around the entire circumference of which have been placed detectors designed to record the energy of the ripples. We would clearly expect them all to detect an exactly equal share of the ripple's energy simultaneously as it spreads out against the tank's circular wall, but the result analogous to our Young's slits experiment is one where just a single detector, at a single location, records the entire energy given to the body of water by the impact of the pebble!

Although the behaviour exhibited by individual electrons is distinctly particle-like, the distribution pattern of large numbers of electrons is entirely indicative of wavelike behaviour. To analyse events more carefully, let us arrange to slow our electron gun down so that it only fires one electron at a time, say one a second. We run the experiment for ten minutes and duly hear a single click at a single counter once every second. But an analysis of the distribution of these 600 clicks again reveals an interference pattern—of *two* wave fronts. Indeed, we could have fired just one electron every hour for 25 days and still have recorded a similar interference pattern. How can this be? How can these electrons be interfering with each other?

The bizarre answer is that each electron must be interfering with *itself*. Each electron must pass through both holes simultaneously and somehow interfere with itself by propagating as two distinct waves which, on interaction with the screen, magically reconstitute themselves in order that the electron be detected as a single point event! You might think it would be possible to confirm this explanation by actually observing an electron passing through both holes, but this proves to be quite impossible. Any attempt to make such an observation will only result in the detection of an electron at either one hole or the other. The electron cannot enter our observational reality in an ambiguous way. By looking for the electron, we force it to 'make up its mind' about exactly where it is. It can only pass through both slits when we are not looking, that is, when it does not have to make any firm decision about its precise position. The conclusion appears to be, then, that reality is only conferred upon the electron in the act of observation—when it is *forced* to commit itself to appearing at one particular point in space.

The mathematics suggests that after passing through the slits the electron continues to propagate as a superposition of two ghost electrons—two separate *ideas* of the electron—each ghost representing one possible state of the 'real' electron. These ghost electrons perform a phantom dance with each other to create a ghostly interference pattern. What is being propagated through our apparatus cannot be regarded as a 'thing' in any conceivable way. We can only think of the electron in very abstract terms as some kind of pattern of information, strictly speaking a pattern of probability distribution. For each moment in time, the mathematics of quantum theory describes the chances of our finding the electron at any particular point of space—*should we look*. It represents a kind of three-dimensional envelope of possibility, a description of the spread of its 'tendency to exist' throughout space.

Because our ghost electron has an equal tendency to pass through both slits, it will continue to propagate as a superposition of these two distinct possibilities. It emerges as two ghost electrons instead of one. Because their patterns of 'tendency to exist' are wavelike in nature, they will interfere with each other to produce an interference pattern of 'tendency to exist'. It is this complex pattern of possibility which governs where the electron will be detected among our array of Geiger counters, not by predicting the exact point at which a click will be registered, but by predicting exactly what *chance* it has of triggering a click at each counter. The mathematics can only talk in terms of likelihoods. We can pick one particular counter and the mathematics will precisely predict the likelihood of an electron registering a click. But it cannot predict whether or not the counter *will* actually click. That is completely unpredictable, indeed, as far as we presently understand, *absolutely* unpredictable.

This unpredictability is not epiphenomenal to the quantum world. The uncertainty in existence which we have just explored is actually enshrined in the logic of quantum theory in the most fundamental way possible. An electron, like any fundamental pattern, simply does not possess a well-defined location and motion. We desperately want to picture it as a 'thing', with a fixed trajectory and ordinary physical attributes, but such concepts have no validity in the quantum world. The image of electrons as particles set in fixed orbits around the nucleus of the atom has now given way to an image of a diffuse cloud that envelops the nucleus, in some places dense, symbolizing high probability; in other places thin, symbolizing low probability. This is the kind of picture that appears in most modern science books. But even this more authentic image is really quite ridiculous. This is a picture of probability! But of course probability is a purely mathematical abstraction. Probability doesn't extend into space. Probability is an *idea*.

How, then, are we to picture the electrons in an atom? Once again, the answer has to be that we can't. As suggested earlier, trying to visualize an electron is, quite literally, as meaningless as trying to smell it. At this level of description, fundamental patterns can only be thought of as patterns of 'potential reality' and can exist in any number of alternative, superposed states. A fundamental probability pattern is not required to be either 'here' *or* 'there'. It can be both 'here' *and* 'there'. These are patterns of 'information', like the software that runs in a computer. Although we can visualize the medium of information storage, whether it be paper, magnetic tape or floppy disk, there is no way in which we can visualize the actual pattern of logic which truly comprises a computer program. This software pattern exists independently of any storage medium, indeed independently of physical reality itself.

The supreme encapsulation of quantum theory's probabilistic foundation is Heisenberg's famous Uncertainty Principle. This principal places a mathematical limit upon the precision with which we can describe the state of any quantum pattern, and is a logical expression of the idea that it is impossible to make a measurement without disturbing and transforming the quality that we are trying to measure. It is quite impossible to look at an electron, like any fundamental pattern, from a detached standpoint. Looking implies a deliberate act of intervention, a participation in reality. In order to know *anything* about an electron, we are compelled to interact with it in some way, thereby irrevocably altering its state. Because of the disturbance of our observation, we can never bring a quantum pattern entirely into focus. This is an absolutely fundamental and profound truth. It represents an epistemological hole from which the physicist has no escape.

But this is not what Heisenberg's uncertainty represents. Heisenberg's uncertainty actually represents an even more fundamental and profound truth, an even deeper epistemological hole, one from which reality itself seems to have no escape. In unavoidably vague and clumsy language, it appears to be an expression of the impossibility of perfect self-knowledge, the need of a fundamental pattern to 'look' outside of itself for a reflective definition. Heisenberg's uncertainty of knowledge is therefore an *intrinsic* uncertainty. It represents an inherent blurring within the very ground of physical reality.

This quantum fuzziness is so pervasive that it even invades empty space, transforming it into a frenzied blur of spontaneous, ceaseless activity. Quantum theory carries on where relativity theory left off in totally debunking our idealization of the concept of the vacuum as a volume of nothingness, portraying it instead as a turbulent sea of transient, acutely ephemeral patterns of energy, continuously bubbling up out of, and settling back into, a seething mathematical froth. Over unimaginably short time-scales, the uncertainty in energy is enough to allow the momentary creation of fundamental patterns out of 'nothing'. This 'stealing of energy as long as it is put back quickly enough' is perfectly legitimate under the uncertainty principle, always remembering that the more flagrant the theft of energy, the less time there is available before it must be returned, and noting that we are talking here about times of the order of a millionth of a millionth of a millionth of a millionth of a second. Although these virtual patterns must almost instantaneously disappear in order to stay 'legitimate', their brief entry on to the stage of reality still enables them to interact with 'real' patterns, an ability which endows them with very great power.

For no better reason than that it is quite astonishing, but also because it so beautifully illustrates the role of quantum theory in providing the kind of creative licence that is necessary if 'interesting' things are to happen in a universe, I want to take a quick look at the phenomenon of radioactivity. In classical terms there is no explanation for the radioactive decay of atomic nuclei. There is absolutely no way in which any of the fundamental patterns responsible for radioactivity can obtain sufficient energy to escape the energy barrier of the nucleus. We can liken the situation to that of rolling a marble up the side of a hill. Speaking classically, if we don't give the marble sufficient energy, it will never get over the crest of the hill. It rolls up the side, cannot quite make the top, and so rolls back down again. No matter how many times we roll the marble with the same speed or energy it will never gain the top of the hill.

In quantum terms, however, the situation is not so hopeless. The marble can do some crafty dealing on the quantum 'black market'. It can borrow some energy against the uncertainty principle and hang on to it just long enough to get over the crest of the hill, by which time it is rolling safely away down the other side and can quite happily repay the loan—all in the twinkle of an eye and without contravening any law of nature! This is just what an alphaparticle does to escape the nucleus of an atom. The greater the energy that needs to be borrowed, the less time there is available to complete the transaction, and hence the less the chance of a successful transaction, the less the chance of radioactive decay occurring in a given period of time. Although it is impossible to predict whether or not a particular nucleus will actually decay in this time, because there are literally millions of millions of millions of atoms in even the most microscopic of samples, we can know with almost supernatural precision what *proportion* will decay. By combining chance with large numbers, quantum theory is able to arrive at near certainty.

It is an amazing fact that the possibility of such briefly transacted 'black-market' activity is at the heart of what we experience as the electric force. The repulsive interaction of two electrons is actually caused by the interaction of their enshrouding clouds of virtual possibility patterns. Indeed, electrons cannot be meaningfully defined without taking account of the shadow troupes of virtual patterns that fleetingly bubble in and out of possibility around them, each in turn pulling along their own shadow troupes of doubly virtual patterns—penumbras of possibility within possibility—and so on, in an infinite nightmare of nested entanglement. It is this maelstrom of instantaneous 'hello-and-goodbye' interaction which actually seems to confer upon the bare electron an apparent extension in space. Rather peculiarly, in order to get solutions out of the mathematics—using a wonderfully outrageous fudge called renormalization by which infinities are cancelled away and replaced by the empirically derived properties of the observed electron (that is, those quantities that have been predetermined by experiment)—it has to be assumed that bare electrons are dimensionless points with a spatial extension of precisely zero. In fact, the physicist's bare electron turns out to be nothing but a mathematical idealization, a mathematical fiction. A bare electron is a construction of convenience, an abstraction that lives only in the minds of scientists.

Fundamental patterns have no existence apart from the self-created context of their troupes of virtual patterns. They are patterns within patterns within patterns, vortices in a whitewater ocean of energy. A fundamental pattern is as inseparable from this sea of virtual energy as a whirlpool is from its sea of water. A whirlpool cannot be isolated and its characteristics examined independently of the ocean. It has no existence outside the context represented by the flux of the surrounding water, a flux participated in by the entire ocean. Remove that support and the whirlpool will simply flow away into a meaningless puddle! Similarly, fundamental patterns are just as meaningless outside the context of the energy flux in which they swim, a flux participated in by the entire universe. Every pattern is recursively blurred into every other against a restless background of frenetic uncertainty. There is no well-defined point at which one pattern ends and another begins. All boundaries dissolve away to leave an irreducible whole. At this level of description, the entire universe is one, a totality of continuously transforming possibility.

Yet out of this ethereal, phantasmagoric underworld of ghostly patterns, out of all this probabilistic subterfuge, the arcane magic of quantum theory appears somehow to conjure up permanence and solidity. Remember, despite all that strange and spooky interference as the electron passed, as a wave of possibility, through both slits, our Geiger counter only ever clicked in an unambiguous response to one perfectly discrete, perfectly real electron. It seems that all the time we are not looking, the wave function—the mathematical representation of our knowledge of a quantum pattern's 'tendency to exist'—is quite happy to live in its own abstract world of limbo. But if we take a 'peek' into this world in some way, the state of our knowledge changes irreversibly. The wave function collapses. Just one actuality precipitates out of a multiplicity of possibility. Our observational intervention into quantum reality exorcises the ghosts!

Our universe appears to possess a schizophrenic split in its character. The personality which we are so familiar with in everyday life disintegrates the moment we get to its heart by way of our most probing experimental questions. We have to admit the existence of two distinct, seemingly irreconcilable worlds: the microscopic world of ghostly quantum patterns and the macroscopic world in which we lead our everyday lives. How are they related? It may come as something of a surprise to learn that physicists do not really have an answer to this rather fundamental question!

We have now arrived at the paradoxical crux of this whole weird business. If we make the standard assumption that our everyday macroscopic world is just a very large collection of microscopic quantum components, at what point do we experience the change from the unending proliferation of ghostly superpositions to the wholly unambiguous reality which we actually observe? Just how does unequivocality crystallize out of the rampantly equivocal quantum underworld? When does the mystical incarnation of just one of these myriad potentialities actually occur? What is the relationship of the whole to the parts? These are all different ways of asking the sixty-four-billion-dollar question of quantum theory. And no truly satisfactory answer has yet been found.

In order to hold some kind of position, the majority of physicists invoke the Copenhagen Interpretation. This rather evasive philosophy was originated by Neils Bohr in 1927 when he introduced the Principle of Complementarity: the wave aspect and the particle aspect of quantum duality are two *complementary* descriptions of one reality, neither of which is complete. Wave features and particle features are mutually exclusive, complementary characteristics exhibited by microscopic quantum phenomena in relation to the macroscopic measuring apparatus necessary to observe them. If we arrange for our apparatus to look for wave features, then waves we shall find. If we arrange for our apparatus to look for particle features, then particles we shall find. But, however we arrange our apparatus, we will never find waves and particles simultaneously.

In this way, quantum theory appears solely as a method of connecting up question with answer. In the Copenhagen Interpretation, it is the act of observation itself which is held responsible for the collapse of the wave-function, the crystallization of actuality out of potentiality. This is how the scientific 'establishment' answers our sixty-four-billion-dollar question—or, more properly, manages to avoid answering it. Armed with this principle and the abstract mathematical tools of their trade, quantum physicists are able to protect themselves against the disorientating effects of quantum magic. They tell us that it is meaningless to talk about any reality other than the reality of our observations. Nothing is to be regarded as 'real' or as having 'existence' until registered by some kind of measuring or observing apparatus.

The Copenhagen Interpretation tells us that we can have no answer without a corresponding question, that no quality of the quantum world has any reality until we can relate to it by way of some kind of observation. The observed and the observation have always to be considered together. We are told that it's a package deal. What makes the Copenhagen Interpretation rather evasive is that no real attempt is made to define exactly what is meant by an act of observation. There is no concern with the question of where it is exactly that the observed becomes the observation. Within this scheme, it is quite sufficient that macroscopic, classical measuring instruments, with large-scale dials and pointers, provide a totally unambiguous indication of quantum measurements. Somehow, somewhere, quantum physics turns conveniently into classical physics and it simply does not matter that it is quite impossible to point to exactly where this abruptly discontinuous transformation actually occurs. It makes no working difference to the typically pragmatic scientist.

Resigning from the need to answer a question does not, however, make that question go away. At what point *does* the metamorphosis of knowledge occur? Where *does* the observed end and the observing begin? There is a considerable body of opinion which suggests that the observing begins in the mind. Human consciousness is held to be responsible for the metamorphosis of knowledge. However, to my mind, throwing one enigmatic concept at another like this seems only to succeed in squaring the number of questions. The difficulties can best be exposed by way of Schrödinger's famous cat paradox.

Schrödinger's notorious cat—a *theoretical* cat—is encased in a sealed box and serves as a macabre form of measuring device. As a result of some quantum event, there is an exactly fifty-fifty chance that an electron will be fired to the left or to the right. If it moves to the left, it will trigger a sequence of events which results in the cat being exposed to a lethal dose of poisonous gas. If it moves to the right, it will trigger a sequence of events which results in the cat being fed its dinner. Immediately following the initiating event, the electron exists as a superposition of a left-moving pattern of 'tendency to exist' and a similar right-moving pattern. According to the mathematics, the electron is not left-moving *or* right-moving. It is left-moving *and* right-moving. Indeed, according to the mathematics, we can theoretically continue this additive description right through the chain of amplifying processes which eventually ends in the cat either being fed or poisoned. And even here, in theory, the mathematics suggests that the cat should be thought of as a superposition of two cats, one very full cat and one very dead cat! As far as the mathematical formalism of quantum theory is concerned, the only difference between an electron and a cat is in the complexity of their probability patterns. Because the box is perfectly sound-proofed, it is impossible to infer our quantum measurement until we actually open the lid to see whether the cat is dead or alive! The question is: does the cat exist in a state of dead *and* alive limbo until we actually open the box, or is it perfectly capable of resolving its lethal identity crisis by itself?

Well, I cannot possibly accept that Schrödinger's poor cat could ever exist in a state of schizophrenic quantum limbo, and I'm sure very few people would try too hard to argue otherwise, yet such is the bottomless nature of the quantum enigma that propositions like these are still discussed by philosophers in all seriousness. What if we were to replace the cat with a rabbit, or a rat, or a fish? What complexity of mind is required to collapse the wave function? Would the mind of an ant suffice, or would it require the collective mind of an entire ant colony? And what about the 'mind' of yet lower forms of life? Where can the line be drawn? What about the question of different states of mind. Do we only collapse wave functions that impinge directly on our immediate consciousness, or can we collapse them subconsciously? The whole question of qualifying mind is a greater philosophical quagmire than quantum theory itself. Also, if one accepts that mind is an evolutionary phenomenon, then the most enormous difficulty of all is encountered, for it surely follows that we must deny the existence of a fixed reality until the very moment of evolution of mind. And how can one possibly imagine such a moment—a moment that marks the point at which some mind becomes sufficiently complex to be able to collapse spontaneously the burgeoning wave function representing the entire universe, thus, somewhat belatedly I would suggest, bringing its very own neurons into existence!

Schrödinger concocted his cat paradox to expose the incompleteness of the Copenhagen Interpretation and succeeded beautifully. The mathematics of quantum theory fits our experimental observations so precisely that its validity cannot be denied, yet still we do not know for sure just what it all means. It is all diabolically tantalizing. If this is your first encounter with these difficult ideas and you are struggling to understand them, don't worry. I have lived with them for many years now and am still struggling to understand them. The most important point to grasp is the very unfathomableness of it all. By virtue of its formal structure, quantum theory has brought us to the edge of an epistemological abyss, a yawning chasm which we are forbidden to span with an objective bridge. In this chapter, then, I have sought to bring you to a position from which you can stare down into this abyss, and actually see for yourself that there is no bottom, that this mysterious well is never going to be plumbed with logic. I have been trying to communicate an understanding of what it is about quantum theory which lies beyond our understanding. If you can comprehend the awesome perplexity unleashed by its sixty-four-billion-dollar conundrum, you have gained access to an exclusive club. You can unashamedly scratch your head alongside the greatest minds of the century!

If any clear message has emerged from our mind-bending peregrination it is that our reality is not all that it seems to be, that the way we normally construe reality—as an impersonal, objective universe containing lots of little personal, subjective universes—is quite wrong. It is this nice, cosy dualistic perception of things, which represents the metaphysical scaffolding of the Western world, that has been most severely undermined by quantum theory. This, basically, is why the physics is so incomprehensible to us. We are foolishly trying to fit a round peg into a square hole. Quantum theory is telling us that we are looking at the world in the wrong way, with the wrong preconceptions. We need to go beyond the standard assumption that our everyday macroscopic world is just a very large collection of microscopic components. We need to invoke a rationality which can embrace the interpenetration of object and subject, a rationality of wholeness to complement the standard scientific rationality of the parts.

Object can no more be separated from subject than the inner surface of a sphere can be separated from its outer surface, or a reflection can be separated from the reflecting surface of a mirror, or a dream can be separated from a sleeping person. This is certainly not to say that there is no subject/object dichotomy, only that the nature of this dichotomy is such that there is always a mutual dependency and interpenetration of subject and object. Quantum theory has demolished the idea that we can talk meaningfully of an entirely objective reality because our knowledge is only ever related to the questions that we ask. These fundamental relations necessarily have their objective and subjective poles, but just as it is impossible to isolate a single magnetic pole, so it is similarly impossible to isolate object or subject.

The message is that the part and the whole, the micro-world and the macro-world, are symbiotically scrambled together. In the words of Paul Davies, 'the macro world needs the micro world to constitute it and the micro world needs the macro world to define it'.[3] Wholes cannot be defined simply in terms of their parts because those parts, in turn, can only be defined in terms of the whole. All the paradoxes thrown up by the quantum measurement problem can be traced back to this same epistemic recursivity, the mysterious feedback of knowledge from the macro-world to the micro-world. Through a higher rationality than that which can be conceived by the human mind, the reality of the objective is somehow interwoven with the reality of the subjective. Classically, we think of mental process as a patterning in material substance. Now we are made to think of material substance as a patterning in mental process. Neither view on its own can be considered correct. Both perspectives are necessary for a complete picture. This is what quantum theory is telling us. The trouble is, once more, that we don't yet know how to understand it.

The rainbow that we see as a glorious arc of colour in the sky does not exist as an object 'out there'. It is a miraculous illusion performed by our alchemist Mind. We know that if a team of physicists took to the air, with any amount of experimental apparatus, they would never be able to measure the widths of those bands of colour in the sky. The idea is, of course, quite absurd. But we have now seen that the objects of our everyday world are really no more tangible than the rainbow. Seemingly objective properties like solidity and extension in space have turned out to be magical symbols of our consciousness. Physicists have searched hard for them and, like the coloured bands of the rainbow, have failed to find them. The closer they have come, the more intangibly elusive these properties have appeared to be. We have witnessed the physicality of the universe dissolve into a symbolic world of mathematical functions and operators, finally to disappear altogether amidst quantum waves of 'tendency to exist'. In pursuit of the ultimate 'stuff' of nature, physicists have been led down a narrow alley, and finally into a dead-end. But, instead of finding their quarry trapped, it is nowhere to be seen. All they are left with is the strange symbolic trace of the shadow cast by their own observational intrusion.

In its quest to discover how the patterns of reality are organized, the story of modern science hints at a picture of a set of nested Chinese puzzle boxes, each one more intricately structured and wondrous than the last. Every time the final box appears to have been reached, a key has been found which has opened up another, revealing a new universe even more breathtakingly improbable in its conception. We are now forced to suspect that, for human reason, there is no last box, that, in some deeply mysterious, virtually unfathomable, self-reflective way, every time we open a still smaller box, we are actually being brought closer to the box with which we started, the box which contains our own conscious experience of the world. This is why no theory of knowledge, no epistemology, can ever escape being consumed by its own self-generated paradoxes. And this is why we must consider the universe to be irredeemably mystical.

The concluding message is that in object, we always see a reflection of subject. This is the concept around which we will have to write the narrative of our new scientific mythology. We need a physics which will bring a sense of involvement to the pattern of our perception, and hence provide a foundation for an enlarged, more authentic, more responsible sense of selfhood. As so perfectly put by Morris Berman, 'We are sensuous participants in the very world we seek to describe.'[4]

3

Exorcising The Ghost of Cartesianism

OVER SIXTY YEARS having now passed since the original mathematical formulation of quantum theory, there are very few
informed scientists around today who would support the absolute
split between subject and object as propounded by Descartes.
Nevertheless, his basic dualistic premise has become so deeply
embedded in the general scientific ethos that, although scientists
now dismiss the simplistic Cartesian dichotomy between mind and
matter, they nevertheless implicitly affirm it in the way that they
mathematically model the world. The scientific pursuit is underpinned by the assumed independence of subject and object, and its
terrific success has allowed scientists to casually overlook the
epistemological frailty of such an assumption. Cartesianism has
certainly been left behind intellectually, but it remains ever present
in spirit. We are still living with its spectral legacy—a ghost that
continues to haunt the modern scientific perception of the world,
a ghost called reason, the rationality of logic and analytical thought.

Of course in its appropriate domain, it cannot be disputed that
reason does reign supreme. It is reason which has brought science
breathtakingly close to making complete its understanding of the
fundamental laws which govern the subatomic parts of the universe,
laws which endow the scientist with an awesome power of
prediction. However, what is invariably overlooked is the fact that
these parts, the ultimate objects of scientific study, have no real
existence. They are mathematical abstractions invented by analytical reason. They are schematic reductions of what is an essentially
organic, irreducible whole. They exist only in the mind. And reason
exists only in the mind. And yet mind is just what reason has
inevitably excluded from its picture of the universe.

By doubting everything, Descartes arrived at just one absolute certainty in his world—his thinking self. Any true system of knowledge had to be centred upon the thinking, reasoning human mind. But we have to remember that Descartes' rigorous doubt was not so rigorous that it extended to questioning the existence of God—who, in Descartes' system, functioned as a guarantor of reason. Since it was a just and good God who had originally granted humanity the tool of reason, it could safely be assumed that God had guaranteed the validity of that tool. Through reason, God had provided humanity with the opportunity to gain complete knowledge of the world, and potentially, so it seemed, complete mastery of the world. The rational intellect became enshrined as the supreme source of knowledge, the only authentic source of true understanding. In this way, Descartes' dualistic system represented the logical climax to a development which started with Plato, the first and arguably the greatest ever apostle of reason. In fact, following in the footsteps of Plato, the whole tradition of Western philosophy can be seen in terms of the growing ascendancy of the rationality of reason over the irrationality of intuition. Finally, once the scientific programme got into full gear, intuitive, experiential knowledge was consigned to the remotest of backwaters. The emotional knowledge of feeling, the symbolic knowledge of dream, the poetic knowledge of myth, all became relegated to a merely incidental status in the great rational vision of the world promoted by science.

In this vision, of course, science found no place for God, and without God, reason was left hanging precariously in mid-air, with no metaphysical means of support. An absolute faith in the irrational divinity of God was replaced by an absolute faith in the rational divinity of reason. But by placing such complete faith in the power of reason, science let fly something of a boomerang—a boomerang which was always destined to come winging its way back into the picture. As we have seen, it eventually returned, low and fast, to scythe out the very ground upon which classical physicists were standing. Indeed, even mathematicians felt the earth cut out from under their feet when, in 1931, Kurt Gödel proved that the very queen of sciences herself could never be made completely watertight, or, more accurately, could never be made watertight in such a way that she could stay afloat. Mathematics can only be made complete at the expense of logical inconsistency in its axiomatic foundations. Through the ingenious metamathematical elaboration of one of the greatest ever intuitive hunches, Gödel was able to prove that mathematics cannot be both consistent and complete. To build a system upon logically consistent foundations is to build a system with an infinite number of leaks. And to build a watertight system is to build a system which is pre-sabotaged with mines of paradox.

In simple terms, the return of the boomerang represented the discovery that self-reference, by transforming object into subject, imposes limits on the rationality of analytical reason. Although this discovery in itself threw up no truly compelling argument for the desanctification of the sacred status of reason, it did serve to throw a bright spotlight on the participating role of the observing mind. In *Quantum Questions*, a work which gathers together the mystical writings of the great physicists of the quantum revolution, Ken Wilber points out that modern physics offers no positive support for a religious world-view, nor does it offer any objection to such a view. It stands indifferent. Yet, as revealed in their writings, all the great pioneers of the new physics—Heisenberg, Schrödinger, Einstein, Eddington, DeBroglie, Planck, Pauli—were inclined toward mysticism. All these men displayed a conspicuous humility before the spiritual mystery. Their imaginations were inspired and also humbled by the epistemological constraints unveiled by the new physics.

As Ken Wilber forcibly emphasizes in *Quantum Questions*, the most truly significant message of twentieth-century physics is that we are not yet in contact with ultimate reality. Not that the new physics has really pointed to any new way forward. Essentially, what it has done is to show us that we have been running down an epistemological cul-de-sac. Indeed, as we have seen, it has exposed the whole Cartesian programme as fraudulent. Science went out in search of the objective components of an entirely material reality, but carrying the Cartesian dichotomy into their experimental battle with nature meant that they were also carrying the seed of their own defeat. In quantum theory we have seen that defeat enacted.

Ken Wilber points out that it was 'this radical failure of physics, and not its supposed similarities to mysticism, that paradoxically led so many physicists to a mystical view of the world'.[1] Wilber goes on to quote from Eddington:

> We have learnt that the exploration of the external world by the methods of physical science leads not to a concrete reality but to a shadow world of symbols, beneath which those methods are unadapted for penetrating. Feeling that there must be more behind, we return to our starting point in *human* consciousness—the one centre where more might become known. There we find other stirrings, other revelations than those conditioned by the world of symbols.[2]

The original and most immediate of these stirrings in our consciousness are clearly those belonging to our senses. It is important in our discussion, therefore, that we briefly consider the nature of our sensory experience of the world. As the most dramatic of our windows on reality, I have chosen vision as my example. Through the unconscious processes of our visual perception, we are provided with a three-dimensional picture of seemingly faultless perfection. Indeed, it is so extraordinary a picture that we can actually live for years without ever bringing into consciousness the fact that our vision is an internal movie, showing inside our own personal cerebral cinema. We are invariably to be caught confusing this movie with the world outside. We are granted a view so astonishingly real that our mental images seem to be mysteriously transcendent. They appear in our mind in such a way that we cannot help but consider them to represent reality as it 'really' is, instead of just our conscious interpretation of the light that has been reflected into our eyes. By imposing this personal interpretation back upon the external world, we often forget ourselves and think that nature abounds with resplendent colour. This perfectly wonderful delusion is almost universal to human perception. For example, it is entirely wrong to talk about the fragrance of the rose as if the fragrance actually belongs to the rose. In fact, it has nothing to do with the rose at all. There is no such 'thing' as a scent which surrounds the bloom, only a vapour of volatile organic molecules. There is no objective chemical property called smell. The scent of the rose is entirely created by *us*, by our alchemist Mind, when the chemical vapour is breathed in and dissolved in the mucus which lines the olfactory cells of the nose.

Let us return to vision, though, the details of which are somewhat easier to think about than smell. The visual process clearly starts with the eye's capture of an array of light waves reflected from the objects in our field of view. This pattern of waves is focused by the lens on to the retina, a layer of two basic kinds of photosensitive cell: rods and cones. The rods are specialized to cope with low-intensity light. The cones operate in normal light and provide for our colour vision in a way that is not yet completely understood. The essential element is the triggering of a photo-chemical reaction within the cone cell and the subsequent genera-tion of an electrical nerve impulse. There appear to be three different types of cone, each with its own characteristic response pattern to the visible spectrum, and each being excited most strongly in just one band of the spectrum. It is by integrating these three specific sources of information that colours are eventually allocated within our visual image.

Leaving the retina, then, is an inordinately complex array of electrical signals, none of which have any direct correspondence to the individual colours that we finally see. These signals are carried along the optic nerve—a bundle of about one million nerve fibres—towards the primary visual cortex—a greyish mass of about 100 million neurons or nerve cells—in a rather less than straightforward manner. The visual cortex, like the brain as a whole, is divided into two halves, each half receiving information from both of the eyes, corresponding to one side of the visual field. The information for the left side gets directed to the right visual cortex whilst that for the right side gets directed to the left. Finally, and quite miraculously, the brain unscrambles this fantastic wealth of electrical data, merging together the two original wave patterns, adding a third dimension of depth, and along the way, and quite wonderfully, conjuring up that magical fourth dimension of colour.

71

It has to represent nature's most mind-blowing piece of technical wizardry. The astonishing end result is that our field of vision appears somehow to transcend the physical confines of our brain. Throughout our everyday life, we really do believe that we see the external world as it actually exists. We are continually fooled by our own brilliant perceptual illusions. We are three-dimensional artists of such awesome talent that we delude ourselves into thinking our creations are real!

Although we can readily admire the fiendishly clever precision engineering of the eye, it is far more difficult to appreciate properly the truly staggering intricacy and complexity of the hidden processes which are actually most responsible for bringing about the miracle of our vision. Contrary to what we might have expected, the data collected by our eyes is not merely copied on to some kind of mental photographic plate. The raw data of the eye is first subjected to a vast amount of computer-type processing to produce the smooth data from which our immaculate visual picture is constructed—in a form which is quite inaccessible to any kind of logical analysis. Indeed, if we try to think about what seeing really is; if we try to conceptualize the origin and nature of our visual imagery; if we try to imagine where these images could possibly be localized, the mind simply boggles.

What we see, then, is not reality, but a breathtaking, techni-coloured, metaphorical model of it created by the dazzling artistry of our consciousness. That we unconsciously impose this model back on reality is an understandable, but absurdly anthropocentric fact of everyday life. Finally, perhaps we should all stand back in awe of our own individual artistic and technical ability. Those exquisite colours of the rainbow are, for each one of us it seems, a very special piece of unique creative magic. *Or are they?* In fact, it is so amazing a feat that we are surely forced to question whether, in fact, that magical palette of colour really can be our own. Can the mind, as understood by science to be purely the product of electrical and molecular activity in the brain, truly be *entirely* responsible for such sublime sorcery?

I look out of my window. One part of the sky is grey and dark. It is raining. Another part of the sky is blue and bright. The sun appears. And so does a rainbow, a diaphanous synthesis of sun and rain, and mind. My mind? Is anyone else seeing a rainbow? Am I the only person bringing 'this' rainbow into reality? I sit transfixed, staring at the colours 'out there', contemplating how it is that they are really 'in here', in my mind. Then I baulk at the sudden realization that all of nature laid out 'before' my eyes is equally as illusory as the rainbow. My attention turns to the rainbow's spectrum of colour. I am both fascinated and frustrated by my hopeless inability to define my chromatic experience and compare it to your own. Although there is no way in which I will ever be able to know for sure, I strongly suspect that our subjective experience of the rainbow's colours is the same. And then, if we do each experience colour in an identical way, what is the nature of the pattern which connects our individual experiences? Where does colour originate? Where is it conceived? I find myself sitting in awe of colour's enigmatic ontology. Since nature is always so conservative, colour seems like an extravagant, almost ridiculous luxury. Why have we been endowed with such an exquisitely beautiful outlook upon the world? Why did nature not settle for 'black and white' vision, which would surely have been perfectly adequate and, technologically, a whole lot easier? Does colour 'merely' appear as an epiphenomenon emerging out of the integrated electrical activity of countless millions of neurons in our brains, or does it perhaps appear by way of some unknown, possibly unknowable pattern, belonging to some higher rationality than that which is presently acknowledged by science? And finally, as perhaps the most unambiguous symbol of our sensory awareness, the questions raised by colour prompt us to ask the most fundamental question: why do we experience anything? Why is our sensory knowledge of the world conscious at all?

If the scientific model of the universe was sufficient unto itself, fully capable of explaining all mental characteristics purely in terms of physical brain activity, there would seem to be no good reason why we should have any internal awareness of our being in the world at all. We already understand that much of

73

what we do in life is the result of unconscious processes. In fact we perform all sorts of activities, from driving to work in the morning to brushing our teeth at night, without the accompaniment of full conscious awareness. It is not difficult, therefore, to stretch the imagination a little further to envisage a world, quite identical to ours in every practical respect, where all human activity is performed unconsciously, trance-like, without awareness. The pseudo-human beings inhabiting this world would be automatons, slavishly following the electrochemical programmes written into their brains, performing the very same tasks that we perform, but totally unaware of themselves and their world. From the outside, there would seem to be no means of distinguishing between this perfectly mechanical world and our own. The pseudo-humans would certainly protest that they were fully conscious, just like us, but their arguments would belie an existence which was devoid of any mindfulness of itself. If we accept the completeness of scientific explanation, these two worlds would be identical in every visible way. All physical laws would be the same. Therefore, by the incontrovertible evidence of our own consciousness, the fact that these worlds are *not* identical is surely overwhelming testimony to the incompleteness of the rationality of science.

A world unaware of itself would scarcely constitute a world at all. Indeed, a world without awareness could hardly even be said to have an existence. Who would it exist for? Despite the conjuring of an illusion of aliveness, it would stand as inert and frigid as stone. Any such world, any such existence without an awareness in which to mirror that existence, would be ridiculously meaningless—which perhaps suggests why our world *is* pervaded by awareness. In the end, there can be no scientific, rational answer to this question. It is a question which unavoidably trespasses into the spiritual domain. My own intuitive feeling is that no such miracle as that represented by the evolution of life on Earth could ever have come about without the co-evolution of some kind of transcendent, immanent Mind. Similarly, I cannot accept the traditional scientific view that our individual conscious awareness is purely epiphenomenal. For me, it is the fundamental principle which actually conceives the very possibility of existence—a spiritual principle, transcending the rationality of science.

The standard Cartesian model of our visual perception, as just presented, considers a separate conscious mind as subject viewing an external world as object. There are three elements involved: the seer, the seeing and the seen—the subject, the experience, and the object. Classical science suggests that the seer and the seen have a concrete material reality, whereas the experience of seeing is very much a kind of added bonus, a curious sideshow to the affairs of the world at large, almost an anomaly. This, however, is very different from the way things seem from our own inner perspective. To us, the *experience* of seeing is the one and only substantial reality. Indeed, the separate experiencing subject and experienced object are both abstractions made *from* the world of experience. Rather than subject meeting object to create the experience of seeing, it would perhaps be better to suggest that the experience of seeing creates the seeing subject and the seen object. Saying this is not to deny the existence of subject and object prior to experience, only to suggest that, outside of experience, there is no meaningful way of defining their existence. Outside of experience, subject and object are undifferentiated within a seamless continuum.

The more I search for the separate self that is the source of my awareness, the source of my seeing, the more I come to suspect that no such self exists. I find seeing, but never the seer. I find hearing, but never the hearer. I find thinking, but never the thinker. I find feelings, but never the feeler. I find memories, but never the remembering self. As Ken Wilber says in *No Boundary:*

> Wherever we look for a self apart from experience, it vanishes *into* experience. When we look for the experiencer, we find only another experience—the subject and object always turn out to be one.[3]

Yet although I can only ever find experience, my sense of self seems to be quite independent of the content of my experience. To begin with, I don't feel that any changes in my physical being would effect my fundamental sense of 'I'. I could undergo all sorts of drastic surgery, completely altering my experience of my own body, but my 'I-ness' would be untouched. Going further, if I were to be deprived of all my senses, I still feel that my 'I-ness' would be preserved. Even if I were to lose all memory of the past events of my life, I still feel that, in some profound sense, I would remain the same essential 'I'.

I can never hope to locate the origin of my 'I-ness' because I can never make my self—the subjective experiencer—the object of my experience. What appears to be my experiencing self, upon closer inspection, always turns out to be an inference, an image captured in some mental mirror, an internal narrative, a projection cast by an awareness of which I can never make myself aware—because it is that awareness which generates my sense of 'I'. My 'I-ness' is independent of the content of my experience because it is created and re-created, moment by moment, *by* my experience, and that experience creates my world too. Here is Ken Wilber again:

> If you are your experiences, you are the world so experienced.
> You do not have a sensation of a bird, you are the sensation
> of a bird. You do not have an experience of a table, you are
> the experience of a table. You do not hear the sound of
> thunder, you are the sound of thunder. The inner sensation
> called 'you' and the outer sensation called 'the world' are one
> and the same sensation. The inner subject and the outer
> object are two names for one feeling, and this is not something
> you *should* feel, it is the only thing you *can* feel.[4]

And the source of this feeling, lacking any better word for it, is *being*. Further, I would like to suggest that the being which creates your self and your world is the very same being which creates my self and my world. And this, fundamentally, is why we agree that our individually created worlds are the same world. Let me requote in part that famous passage of Eddington's from the last chapter:

> In the world of physics we watch a shadowgraph performance
> of the drama of familiar life. The shadow of my elbow rests
> on the shadow table as the shadow ink flows over the shadow

paper. It is all symbolic, and as a symbol the physicist leaves it. Then comes the alchemist Mind who transmutes the symbols. The sparsely spread nuclei of electric force become a tangible solid; their restless agitation becomes the warmth of summer; the octave of aethereal vibrations becomes a gorgeous rainbow.[5]

The error that each one of us naturally makes is to think that Eddington's alchemist Mind is localized in our own head. But to go looking for the origin of the alchemy of being in the internal functioning of the brain represents the same order of mistake as looking for the origin of a television picture in the internal functioning of the set. There can be displayed countless television pictures, but behind them all sits just the one transmission. Similarly, there can be countless individual minds, but behind them all sits just the one alchemist, just the one Mind, and, in the words of Erwin Schrödinger, 'The plurality that we perceive is only *an appearance; it is not real.*'[6]

In the development of my thought and feeling in this direction, there has been no greater influence than Schrödinger. Because he was one of the central figures in the development of quantum theory I couldn't help but have a very great respect for what he had to say. Making my first tentative steps from the scientific domain to the spiritual, his writing seemed to speak to me in an especially personal way. He seemed to offer a means of bridging the gap between these two modes of understanding the world. For that reason, I would here like to quote in full the passage which did most to stir my nascent spiritual imagination. It stands among the finest and most evocative of all contemporary mystical writing.

Suppose you are sitting on a bench beside a path in high mountain country. There are grassy slopes all around, with rocks thrusting through them; on the opposite slope of the valley there is a stretch of scree with a low growth of alder bushes. Woods climb steeply on both sides of the valley, up to the line of treeless pasture; facing you, soaring up from the depths of the valley, is the mighty, glacier-tipped peak, its smooth snowfields and hard-edged rock faces touched at this moment with soft rose colour by the last rays of the departing sun, all marvellously sharp against the clear, pale, transparent blue of the sky.

According to our usual way of looking at it, everything that you are seeing has, apart from small changes, been there for thousands of years before you. After a while—not long—you will no longer exist, and the woods and rocks and sky will continue, unchanged, for thousands of years after you.

What is it that has called you so suddenly out of nothingness to enjoy for a brief while a spectacle which remains quite indifferent to you? The conditions for your existence are almost as old as the rocks. For thousands of years men have striven and suffered and begotten and women have brought forth in pain. A hundred years ago, perhaps, another man sat on this spot; like you, he gazed with awe and yearning in his heart at the dying light on the glaciers. Like you, he was begotten of man and born of woman. He felt brief pain and joy as you do. *Was* he someone else? Was it not you yourself? What is this Self of yours? What was the necessary condition for making the thing conceived this time into *you*, just *you*, and not someone else? What clearly intelligible *scientific* meaning can this 'someone else' really have? If she who is now your mother had cohabited with someone else and had a son by him, and your father had done likewise, would *you* have come to be? Or were you living in them, and in your father's father, thousands of years ago? And even if this is so, why are you not your brother, why is your brother not you, why are you not one of your distant cousins? What justifies you in obstinately discovering this difference—the difference between you and someone else—when objectively what is there is *the same*?

Looking and thinking in that manner you may suddenly come to see, in a flash, the profound rightness of the basic conviction in Vedanta: it is not possible that this unity of knowledge, feeling, and choice which you call *your own* should have sprung into being from nothingness at a given moment not so long ago; rather this knowledge, feeling, and choice are essentially eternal and unchangeable and numerically *one* in all men, nay in all sensitive beings. But not in *this* sense—that *you* are a part, a piece, of an eternal, infinite being, an aspect or modification of it, as in Spinoza's pantheism. For we should then have the same baffling question: which part, which aspect are *you*? What, objectively, differentiates it from the others? No, but, inconceivable as it seems to ordinary reason, you—and all other conscious beings as such—are all in all. Hence this life of yours which you are living is not merely a piece of the entire existence, but is, in a certain sense, the *whole*; only this whole is not so constituted that it can be surveyed in one single glance. This, as we know, is what the Brahmins express in that sacred, mystic formula which is yet really so simple and clear: *Tat tvam asi*, this is you. Or, again, in such words as 'I am in the east and in the west, I am below and above, *I am this whole world*.'

Thus you can throw yourself flat on the ground, stretched out upon Mother Earth, with the certain conviction that you are one with her and she with you. You are as firmly established, as invulnerable, as she—indeed, a thousand times firmer and more invulnerable. As surely as she will engulf you tomorrow, so surely will she bring you forth anew to new striving and suffering. And not merely, 'some day': now, today, every day she is bringing you forth, not *once*, but thousands upon thousands of times, just as every day she engulfs you a thousand times over. For eternally and always there is only *now*, one and the same now; the present is the only thing that has no end.[7]

I must admit to being rather ambivalent about the idea that I share in the same being as yourself. Intuitively, it *feels* right, and is wonderfully appealing, but rationally, it makes no logical sense. Sometimes it seems to represent the most natural of wisdoms; at other times, though, it seems quite patently nonsense. It strikes me then as foolish to argue that I exist as anything other than a separate subject, 'in here', confronting an external world as object, 'out there'. In such a frame of mind, I have to push my intellectual logic to its furthest extremes, and into paradox, before I can recover my intuitive vision. This is the route to intuition that we have pursued in this first section of the book: taking logic to its breaking point— and pushing beyond. I felt that this rather brutal approach was called for in order to counter the arrogance of our logical perception, which, in the modern Western mind, has come to dominate our intuitive perception almost totally. Analysing a whole, in terms of its parts is a very powerful and useful way of looking at the world, but, as we have learnt from science itself, it has vital limitations. The model of the world it creates leaves no room for the ideational knife which cuts that world up into pieces in the first place. A rationality of fragmentation has to be augmented by a rationality of integration, a way of understanding reality in terms of wholes rather than parts.

We have seen that the limitations of scientific observation are the limitations of scientific questioning. The extent of our knowledge is always determined by the kinds of question that we are *able* to ask. Indeed, the story of science has been very much the story of developing increasingly ingenious and more penetrating means of questioning nature. All scientific apparatus, from the simplest example of the humble telescope to the latest generation of inconceivably complex particle accelerators, can be considered an extension of our senses, enabling us to 'see' out to the far reaches of the universe, and 'see' into the innermost reaches of the atom. By fantastically increasing our sensitivity to the deeper patterns of nature, we have discovered that observation is necessarily participation. We must actually interfere with the 'individual' patterns of nature in order to know anything at all about them, with the result that our knowledge is context-dependent, determined by the nature of our interference, the kind of question we have asked.

Just as we cannot see radio waves without the aid of technology, because our eyes are not physically constructed in a way which allows them to interact with the patterns of radio waves, is it not possible that there are patterns which are similarly invisible to the eyes of science? Just as we cannot observe radio waves because our natural senses provide us with no means of asking an appropriate question, is it not possible, even probable, that there are patterns in nature, organizational principles, connections, which science, by virtue of the limited kinds of question that *it* is able to ask, is also disallowed from observing directly? In fact, I would like to suggest that the collapsing wave function provides an obvious case in point. Here is a natural phenomenon which science has been quite unable to embrace with its logical rationality. Here is a qualitative, software aspect of nature which reason has not been able to describe.

Our investigations into the quantum underworld have shown us that observation somehow crystallizes our observed reality out of an indeterminate solution of possibility. Somewhere between the observer and the observed lies a discontinuity, the point of collapse of the wave function, the point at which the observer ends and the observed begins. This intimate interdependency of the worlds of the observer and the observed, tied together in the act of observation, is the Gordian knot of epistemology. At the moment, we are not even close to being able to loosen this knot, let alone to unravel it. One obstacle in particular stands in the way of progress: our lack of a language in which to frame an understanding of wholeness. We have no available means of expressing the qualitative continuity of the three worlds which form our living reality: the world of ideas, the world of experience and sensation, and the world of things. We now understand that the world of things is no more or no less tangible than the world of ideas, that the only boundary between our inner and outer realities is an organizational one, yet we possess no language with which to make this understanding concrete. Worse still, we currently have no consensus which even finds a need for such a language. Within scientific and philosophical circles alike, it is not yet generally recognized that the failure of Cartesian dualism necessarily implies our adoption of some kind of participatory, organic epistemology, an epistemology which can accommodate whole-knowing, the kind of knowing which we normally think of as intuitive.

In emulation of science, the modern philosophical desire for certainty has led to an exactness of description which is exasperatingly tedious in its content, and lacking in anything even vaguely significant to the problems of the human condition.

Academic philosophy has always tended to be remote, but it has now become a joke, an increasingly irrelevant parody of academic science, focused ever more narrowly and obscurely on linguistic minutiae. In their efforts to leave absolutely no room for doubt, modern philosophers have left no room for wisdom either. Along with scientists, they have failed to face up to the fact that their ideal of a strictly logical system of description is a fiction. As Morris Berman states in *The Reenchantment of the World*:

> The Cartesian paradigm is a fraud: there *is* no such thing as purely discursive knowing, and the sickness of our time is not the absence of participation but the stubborn denial that it exists—the denial of the body and its role in our cognition of reality.[8]

As we have already seen, the whole Western tradition of science and philosophy has been built upon reason, with its single foundation in our rational, thinking mind. The unconscious, intuitive mind has been almost entirely ignored, chiefly because it has no distinct voice of its own. The unconscious mind is unable to speak up for itself. This, primarily, is why thought has come to hold such a supreme position. Here is Colin Wilson, from *Beyond The Outsider*:

> This analogy with astronomy perhaps explains the nature of the Cartesian fallacy most clearly. It seems both simple and common sensible to believe that the earth stands still and the heavens revolve; but when astronomers tried to explain the motions of the heavens on this principle, they discovered that it led to complications that defeated them. It seems simple and obvious to assume that the universe will finally be understood if the mind looks on in an attitude of scientific enquiry. *But making the 'I think' the centre of gravity of philosophy is like making the earth the centre of the universe.*[9]

Thinking is always preceded by feeling. There can be no thought without prior felt experience. Indeed, our rational understanding is always to be found embedded in the irrational matrix of experience. In reality, the intellectual knowledge of the 'I that thinks' can never be divorced from the visceral knowledge of the 'I that feels'. As it happens, this intimation that we are home to both a thinking 'I' *and* a feeling 'I' turns out to be not very far from the truth. Thinking and feeling, it seems, are the products of two more or less separate minds.

When we looked at the process of vision a little earlier, I briefly mentioned the fact that the brain is divided into two halves, or hemispheres. It was discovered in the early 1960s, through work led by Roger Sperry and Joseph Bogen, that certain severely epileptic patients could be helped by disconnecting the two halves of their cerebral cortex, achieved by cutting the corpus callosum, a large bundle of nerve fibres which connect the two hemispheres. Somewhat surprisingly, this extreme surgical measure left patients, superficially at least, scarcely changed. They reported feeling no sense of loss and were apparently unaware of any impairment to their mental function. However, closer examination revealed some very bizarre side-effects. If an image was presented to the left visual field, processed by the right hemisphere, the split-brain subject reported seeing nothing. Their visual experience was felt to be complete, yet they could not say that they had seen anything. Nothing seemed to be wrong with their vision, yet verbally subjects reported that no image had been observed. However, if asked to draw what they had seen, they were able, to their own astonishment and great consternation, to recall the image as a picture, with their hands. If different images were offered to the left and right visual fields, the subjects said that they saw one thing, yet drew something completely different. When confronted with this contradiction, they became thoroughly confused. They could not understand why they had drawn a different image because, quite literally, their left brain did not know what their right brain was doing.

These experiments have pointed very clearly to both the autonomy of the two hemispheres and their high degree of specialization. The left brain seems to be specialized for language, the right brain for kinesthetics. If the knowledge of an image is held in the right hemisphere of a split-brain subject, it cannot be expressed in words because, with the corpus callosum severed, that knowledge cannot be accessed by the left hemisphere. The right hemisphere, far more closely linked to the body and visuo-spatial co-ordination, is only able to release this knowledge through the hands. The conclusion to be drawn is that the left brain is home to our rational mind and the right brain to our intuitive mind. Further, there is no good reason to suppose that these two minds are any less autonomous in ourselves than they were found to be in split-brain subjects. The autonomy was simply made very much more apparent in the latter case because the interface between the two minds had been removed.

The rational mind, then, is our communication centre, our language base, the processing centre for our verbal thinking, and therefore the home of our reflective consciousness. Indeed, language has to be considered the conceiving agent of our selfhood, giving birth to our self-aware, self-conscious 'I' by providing the thread which continually weaves together our past history with the present moment. It could be said that language threads together, through memory and imagination, the 'I' of the past, the 'I' of the moment, and the 'I' of our dreams. It is language which enables us to turn thought back on itself and so transform subject into object. It is language which enables us to build the self-image which is our ego, forming the very foundation of our identity as a separate self.

The rational mind, specialized for language and verbal thought, seems to operate analytically, much like a computer, following the digital laws of logic. The intuitive mind, however, is analogic and therefore considerably more difficult to describe. Its most important feature, one which cannot be modelled in logical terms, is its ability to grasp things whole. The intuitive mind is capable of dealing in conceptual wholenesses. It spies pattern or form, shapes in our visual field, sounds in our aural field, meaning in our linguistic field. Intuitive understanding is indeed more of a feeling than a knowing, paying attention more to context than detail. It is an understanding which always appears fuzzy at the edges because our rational mind can never analyse it with its logic. All our deepest and most profound wisdom falls into this category. If we consider our understanding of colour, space and time, beauty, love, or understanding itself, it becomes apparent that all these fundamental concepts are essentially experiential, repulsing rational, verbal thought. They cannot legitimately be reduced. They can only be ingested whole, by our intuition.

As I have already suggested, our intuitive mind is closely attuned with our body. Whereas the rational mind learns through the medium of language, the intuitive mind learns through the medium of experience—in fact most of the knowledge that we employ in daily life is acquired this way. Intuitive knowledge is absorbed, unconsciously, through the body. All our basic, taken-for-granted skills, things like riding a bicycle for example, are assimilated through experience rather than through the intellect, through doing with the body rather than through thinking with the rational mind. If we consider, say, the skills employed in spoken communication, we find that much of our understanding of the meaning of messages is derived not from the actual words that people speak, but from the way that they are spoken, the intonations in the voice and a whole complex, unspoken vocabulary of body language. There are no manuals available for this tacit language, yet, without ever having had to think about it consciously, we all understand its grammar. Throughout our lifetime of experience, the knowledge has been digested by our body and programmed into our unconscious.

In order to move toward a more authentic, organic epistemology, we need to shift the centre of gravity of our philosophy from the abstracted 'I think' toward the bodily 'I feel'. This in turn requires a radical shift in our understanding of mind. Leading the way in this direction has been Gregory Bateson, probably the greatest polymath of the twentieth century. After a career which ranged across biology, anthropology, ethnology, learning theory, psychology, cybernetics and systems theory, Bateson ended his rich intellectual life struggling to outline an epistemology which united fact with value. He was moving toward what he termed an epistemology of the sacred. Unfortunately, it is well beyond the scope of this book to elaborate upon his thought in any detail, for his ideas are both difficult and profound, and cannot be done justice in brief summary. For the purposes of this chapter, we shall have to confine ourselves to just a few principal points.

First, Bateson started out from the thesis that logic and quantity are entirely inappropriate devices for the description of organic systems. He understood that a logical chain of thought, when turned back on itself to create a circuit, is destroyed by paradox. He also understood that it is information, in the form of difference, rather than quantity which is the animating 'stuff' of organic systems. His vision saw through to the metaphoricality of the living world. Bateson argued that mind is not separate from matter, nor is it inherent in matter, but it is immanent within the *organization* of certain patterns in reality. He cited six specific criteria which must be met by a system in order for it to qualify as a mind. For our purposes, the key criteria are: that mind is an aggregate of interacting parts; that the interaction between parts of mind is triggered by difference, with the triggered effect being a coded version or a representation of that difference, in the same way that a map is a coded representation of the mapped territory; and that the system contains circular and more complex data pathways or feedback loops.

By these criteria for mental process it is nonsensical to talk about the localization of our mind within our brain. Under the Batesonian definition of mind, there is no longer any mind/body distinction. The mind is contiguous with the body. But in fact we have to go further than this. It is actually nonsensical even to talk about the localization of mind within our body. Bateson uses the example of a man chopping down a tree. Within the Cartesian paradigm, we would attribute mind to the man and consider him to be wielding an objective, external axe against an objective, external tree. Within the Batesonian paradigm, however, mind cannot be attributed to any *part* of the system. It is immanent within the whole. In cutting down the tree, the man's mind becomes a function of the flux of information which is flowing circularly around the man-axe-tree system. The man, the axe, the tree, the action of chopping, are all patternings within the one encompassing mental process. This whole system of interaction has to be understood to be alive with Mind. According to Bateson's, *Steps to an Ecology of Mind*:

86

The individual mind is immanent but not only in the body. It is immanent also in the pathways and messages outside the body; and there is a larger Mind of which the individual mind is only a subsystem. This larger Mind is comparable to God and is perhaps what some people mean by 'God', but it is still immanent in the total interconnected social system and planetary ecology.[10]

As his final criterion for mind, Bateson requires that a system discloses a hierarchy of levels of representation, or logical typing, with messages at each successive level coding the context of the messages at the preceding level. He is emphasizing here the ability to discriminate between communication and communication about communication, the tacit metacommunication without which a message is meaningless. 'Thank you very much', for example, when said with the appropriate intonation and facial expression, can convey a message virtually opposite in meaning to the literal meaning of the words spoken. Closely related to this is the ability to discriminate between the literal and the metaphoric. The map is not the territory, just as the class of all oranges is not an orange, or the 'man in the street' is not a man at all, but an abstract class of people. These pairs are all examples of different logical types, different levels of representation, and it is vital for the integrity of any system that this difference is recognized—the context of the message must first be understood in order to understand the content of the message—and that there is no mix-up among levels. In fact, it is just such a mix-up, what amounts to a learned disability to discriminate between logical types, between message and context, metaphor and reality, which Bateson held to be the cause of the disintegration of personality exhibited by schizophrenics.

Minds become capable of interesting behaviour, like learning and creativity, when there is two-way communication between the high- and low-order levels of representation, in other words, when the organization of the system is such that representations or symbols at the top level have an influence on those at the bottom level. In our own mind, at the level of conscious awareness, we have very complex symbols which represent such ideas as moral

87

principle and ambition, but beneath this top level there must be many intermediate levels of representation, going right down to the scale of the neuronal network and the molecular biochemistry of the body. Science has been relatively successful in tracing causal chains of determination from bottom to top, isolating causes at the biochemical level and associating them with effects at the level of consciousness, but it has been singularly unsuccessful in tracing causal chains going the other way, from top to bottom. In fact, until quite recently, the reality of such chains of determination was scarcely even recognized. Biochemistry was understood to influence mental state, but mental state was not understood to be able to influence biochemistry—because, basically, whereas biochemistry could be modelled, mental state could not. The traditional scientific picture of things allows for only the one direction of epistemological flow, with the contents of consciousness sitting at the very top, almost superfluously, like the icing on a cake. Conscious thought could not exert a causal influence over the body because the mind was supposed only to be a by-product of the brain.

The Batesonian view of mind tells us that this is rubbish. And we should all know that it is rubbish from experience. We all understand that worry, for instance, can make us ill. The effectiveness of our immune system is closely related to our mental attitude. When we are heavily committed to a project, enjoying our work, buoyant with meaning, not under too much stress, we seem able to keep infection completely at bay. However, when the project finishes, flushed with tiredness and anti-climax, or, alternatively, if the stress has become too great, if pleasure has given way to pressure, we can allow ourselves to become ill. Indeed, the psychosomatic origin of illness must now be suspected to be the rule rather than the exception. The placebo effect offers the classic example of downward causation. The intangible expectation of feeling better can very effectively alleviate physical symptoms. Conversely, a fear or an anxiety—which, again, is nothing more tangible than ideas—can bring a person out in a rash, or induce an ulcer. These examples of psychosomatic healing and illness both offer an unambiguous illustration of the interpenetration of our so-called objective and subjective realities. The ulcer and its causal anxiety are both symbols, of a different class, in the one mental process. Neither is any more or any less concrete than the other.

In our mind, bottom-up and top-down causal chains are always to be found together, chasing each other around a recursively structured hierarchy of representational levels. And it is this exquisitely entangled circularity of mental process which excludes the possibility of a reductionist analysis. Once there is feedback, once statements about a system become mirrored inside the system, a reduction by logic is no longer possible. In this sense, at least, we can confidently say that psychology can never be reduced to physics. A circular, self-reflective system can have no linear, digital mapping. It is a whole, a territory which cannot, in principle, be mapped.

This also points to why intuition is so elusive. Intuitive understanding is a systemic function which involves the interaction of symbols and the relaying of messages at representational levels below the highest, below the level at which consciousness operates. Intuitive understanding is a communication between sub-systems of a mind, like the tacit communication in body language which accompanies a spoken conversation between two people. The understanding inheres in the tangled web of interacting symbols chasing each other around the combined mind of which the two individual minds are sub-minds. This understanding certainly influences the high level symbols at the conscious level, but for the most part there is no mapping into the domain of our rational understanding.

A good example of this is offered by my own understanding of a typewriter keyboard. I spend many hours a day typing away at the keys of my computer, yet I have no rational understanding of the position of each letter. If I quickly try to think about where, say, the 'C' is positioned, my mind clouds over. I cannot locate it. The understanding is in my hands, not in my head—or, more properly, it is in the relationship between my fingers and the keyboard. There is no mapping of my intuitive understanding of the keyboard into symbols accessible by the rational mind. Such an understanding cannot be directly mapped into the rational mind. I can of course learn rationally the positions of the letters, but that knowledge would exist quite separately from my intuitive, functional knowledge of the keyboard.

In search of a little more insight into the nature of our intuitive understanding and its relationship to thought, I would like briefly to discuss the curious concept that we call confidence. A virtuoso ballerina is performing superbly well. We say that she is dancing with tremendous confidence. If, however, she were suddenly to falter and start to make mistakes, we would probably say that she had lost her confidence. What then is this strange quality called confidence which is so ephemeral and so obviously pertinent to artistic excellence? It is far from easy to pin the answer down. From my own experience, the best illustration is to be found in the game of snooker. At the highest level, confidence is the key factor determining quality of performance. This is not quite the case at my relatively mediocre level of play, but it does still remain a very important influence. I play a few good shots and suddenly I 'get' some confidence. I start to play really well, feeling almost as though I could pot any ball on the table. Very occasionally—far too occasionally, unfortunately—I get so confident about a shot that, with an inexplicably strong conviction, I 'know' I am going to pot the ball, sometimes a very difficult ball even, and invariably I do! Then something happens; I play a couple of bad shots, and suddenly I 'lose' all my previous confidence. Now I can't pot even the easiest of balls. Since my potential capacity to play the game surely remains constant throughout these swings in performance, what is it that is actually gained and lost? I would like to suggest that the answer is to be found in the relationship between the left- and right-handed minds. Confidence is a function of the degree of co-operation between the rational mind and the intuitive mind.

Once the necessary intuitive motor skills have been acquired, through many thousands of hours of practice, the best professional snooker players are perfectly capable of potting any ball on the table. At the top of their game, playing with great confidence, relaxed, they are almost oblivious to the mechanics of the cueing action and the basics of the game upon which players of my standard have to concentrate so hard. The rational mind makes the strategic calculations as to where the cue ball needs to move on the table and how that position is to be achieved. The intuitive mind is then trusted to look after the task of actually getting it there, and of course potting the object ball. The precise touch of the cue on the cue ball—the pace with which the cue ball is struck and the kind and amount of spin imparted—is under unconscious rather than conscious control. Ideally, the conscious *logic* of the rational mind is working in tandem with the unconscious *feel* of the intuitive mind. However, if at the end of a match a professional player finds himself in a situation where the entire championship hinges on the last black ball, he is likely to become only too conscious of the feel of the cue in his hands. We watch him take his time over his final, vital shot. We watch him get tense and nervous and lose touch with his intuitive mind, breaking the spell of confidence. After a dazzling show at the table, we watch him muff the match-winning shot. Gripped by self-consciousness, the close collaboration of reason and intuition breaks down as a consequence of the thinking mind's desire to take over conscious control. Not sufficiently trusting of its unassertive partner, the rational consciousness tightened its attention, interrupting the flow around the player-cue-ball circuit of intuitive understanding.

It is one of the less fortunate facts of life that this vital harmony between our left- and right-handed minds is usually at its most elusive just at the time when we most need it. Whatever kind of activity we are engaged in, we perform at our best in a state of relaxed concentration, or concentrated relaxation. It is just this kind of mental state which many techniques of meditation seek to induce. The difficulty for us is that concentration tends to oppose relaxation. We cannot strive to relax. Any conscious *effort* to relax tends inevitably to be self-defeating. I find that I can best attain this state of mind through running. As long as I keep the pace down, so that I am not under any great physical strain, the rhythm of running seems to quieten the rational mind in a focused, concentrated sort of way. This is often the time when I find 'inspiration', when I do my most creative thinking, when I tend to feel connections between previously isolated and unassociated ideas. Strangely, though, it is vital that I commit these thoughts to paper as quickly as possible, for, like the content of dreams, they are despairingly evanescent. They seem to evaporate as I drift back into my everyday, stationary consciousness. I would like to suggest then that creative inspiration, certainly in writing, is another synergetic function of the collaborating rational and intuitive minds. Intuitive understanding is empathetic, an experiential participation in an activity or an object or, most importantly in this context, a concept. But a conceptual intuitive understanding will always remain vague and shadowy, no more than a 'feeling', without some kind of accompanying rational understanding, or interpretation, to function as a framework. The presence of analytical reason is vital to give useful, tangible shape to our conceptual intuitions. Inspiration is a breakthrough, a sudden insight into the nature of an intuition, a partial mapping into a rational, consciously accessible symbol, brought about through the attunement of the rational and intuitive minds.

If anything has become clear from this difficult discussion it should be that mind and consciousness are two quite distinct phenomena. Consciousness is a very special quality which is attributable only to a highly specialized kind of mind—specialized in a way which at present we do not understand very well. There are no specific criteria, analogous to Bateson's, which we can apply to a mind in order to check for consciousness. The human mind, as we can all testify, is a conscious mind, but it is not *just* the conscious mind. Just as the bulk of an iceberg sits submerged beneath the surface of the sea, so the bulk of the human mind sits submerged beneath the surface of consciousness. The human mind is a system which incorporates as sub-systems a number of different unconscious minds. As you may already have appreciated, I have been using the intuitive mind as something of a blanket label to cover all such sub-minds. Some of these, in a way which is again reminiscent of a giant iceberg, reach down into unbelievably strange, seemingly unfathomable depths.

Our dream mind is the obvious example. The extraordinary creativity revealed in the surreal imagery of our dream world totally defies rational explanation. It is wondrously inexplicable. But perhaps most remarkable of all is the correlation of dream symbols across individual people and cultures and races, as revealed most comprehensively in the work of Carl Gustav Jung. Following many years of dream analysis, covering thousands of patients, he found that in order to explain the regular recurrence of certain motifs, often coincident with those appearing in ancient mythologies quite unknown to his patients, it was necessary to postulate the existence of some kind of collective human unconscious. Just as we all share the same basic physical anatomy, so we all share the same basic psychic anatomy, a common instinctual imagination which is home to certain global symbols that Jung called archetypes. If his ideas are valid, and the evidence, although not readily amenable to analysis, is impressive, we seem to be brought full circle. Just as the tip of every conscious human mind appears radiant with the same awareness of being, so the deepest root of every human unconscious appears to be planted in the same symbolic soil.

At both these common nodes we find the suggestion of connections which lie beyond the logical rationality of reason, an adumbration of a mysterious epistemic circuitry within some greater Mind, a Mind in which mental process and material substance exist as internested patternings at different levels of representation. As indicated in the title of his most widely influential book, *Mind and Nature: a necessary unity*, Bateson held the world in which we live to be a Mind in a very real sense. Nature is a mental process, which, it must be very clearly stated, in no way implies that nature possesses consciousness. It should also be added that, within this context, there is much room for discussion as to the nature and extent of the feedback which is possible within such a Universal Mind, and Bateson himself, an extremely careful and rigorous scientist, would very likely have distanced himself from many of the ideas that his work has quite recently spawned—quite possibly including some of the ideas presented here.

Leaving aside the finer details, though, Bateson's conception of Mind provides a new context in which to understand the universe of our experience. To speak in the religious terms that were introduced in the passage from *Steps to an Ecology of Mind*, the human mind is a sub-mind in the Mind of God. This is the idea that I wish to carry tentatively into the second section of this book.

93

And we should be awed by it. The scientific discoveries relating to the low-level order of the Mind of the universe—loosely speaking, the hardware—have been fantastic. But even more fantastic still is what hasn't been discovered, in the nature of the high-level order— the software organization—the immeasurable patterns of down-ward causation from the wholes to their parts; and ultimately the numinous pattern of being itself, alchemist Mind, transforming symbols into sensation, conceiving our universe of experience.

This chapter has inevitably been rather imprecise. It has consisted far more of hints and nudges than clear-cut statements. In fact it has proved impossible to present the ideas in any other way. My main aim has been to open up a fresh perspective on the world, one which has hopefully brought the material world a little more magically to life. To exorcise the ghost of Cartesianism is to transform our perception, to move our epistemological centre to somewhere between thinking and feeling, to begin to live in Morris Berman's enchanted world of sensuous participation. Since his influence on my own thought has been so considerable, I would like to give Berman the last say in this section:

> What I mean by 'Mind' is the conjunction of the world and the body—*all* of the body, brain and ego functions included. Once Mind so defined is recognized as the way we confront the world, we realize that we no longer 'confront' it. Like the alchemist, we permeate it, for we recognize that we are continuous with it. Only a disembodied intellect can confront 'matter', 'data', or 'phenomena'—loaded terms that Western culture uses to maintain the subject/object distinction. With this latter paradigm discarded, we enter the world of sensual science, and leave Descartes behind once and for all.[11]

4

The Mythology of Religion

AS WELL AS SCIENCE, religion is also a major victim of the cult of trivialization. In fact, largely because of its longer history, it is a far worse victim. The original meaning of religion has been almost entirely lost to cliché. As with science, the art fails to be distinguished from the ideology. The excess eschatological baggage of religion has come to obscure its essential spirituality. The poetry of religion has become hidden within its dogma.

At its heart, a religion is a mythology, a way of understanding the meaning of reality and human existence, and, when it is authentic, that understanding has always been communicated through the language of myth. In its attempt to grasp the metaphysical, the language of religion has unavoidably to be symbolic and metaphoric in character. It must speak to our intuition. The trouble is that instead of being read allegorically, myth is open to being read literally, in a naive way, so that its symbolic representations get taken at face value. And this, historically, is just what has tended to happen, much aided and abetted by the institutionalized power-bases of the great religions, whose political best interests have always been well served by such a corrupting process of objectification. Over time, metaphor has invariably become ossified into metaphysics. The myth of religion, its poetry, has become canonized into ideology.

To the critically minded impartial observer, it is unassailably obvious that the great religious myths of the world have been born out of the human psyche rather than out of actual events of the past, yet throughout the world countless millions of people still hold to the literal truth of the myths of their inherited culture. The simple fact is though that the same central themes of mythic thought are to be found in human cultures throughout the world and throughout history. The same motifs can be found in the very similar legends that lie at the ancient roots of many different religions with entirely different cultural origins. Clearly, these myths are not to be traced back to historical fact. According to the great mythologist Joseph Campbell, 'they speak, therefore, not of outside events but of themes of the imagination'.[1] The recurring motifs of myth can be identified with Jung's archetypes, the symbols of the collective unconscious, belonging to the shared imagination of our common biological origin. Myths give poetic expression to the primal impulses of our instinctive nature, the potentialities and motivations which are universal to humankind. They are the distillation of the dreams of the whole human world, addressing timeless, universally human problems. Collectively, they represent an iconography of the soul.

Campbell regards mythic symbols and images as 'messages to the conscious mind from quarters of the spirit unknown to normal daylight consciousness'[2], and as such they must not be regarded as referring to actual events of history. However, over the years these archetypal messages have indeed been misread in just this way, with the result that their power has been deflated, to be backed up in the dead-end of dogma. Jung believed that we possess a natural spiritual instinct which demands expression in much the same way as the sexual instinct, and suggested that our psychic health depends on the development of an authentic outlet for its associated energy. Unfortunately, rather than focusing it creatively, organized religion has generally served to dissipate this spiritual energy, filtering it harmlessly away through the mechanism of devotion and worship to an external deity. Additionally, many of the spiritual symbols of traditional religion have become endowed with an objective as opposed to a subjective function, often leading to a gross perversion of their original value and meaning. It could be argued, in fact, that as a result of this process, many organized religions have become divorced from spirituality. Indeed, in some instances the suggestion must be that the marriage was never properly consummated in the first place.

The great religions of the world have emerged out of vastly different cultural and historical contexts. They have developed under the influence of different value systems and power structures, being designed to meet different social needs. And design is the important word. Our great religious systems were once, and in many areas of the world still are, completely integrated with the socio-political system. They have always had to be relevant to both the needs of the people and the needs of society. During their formative periods their doctrine was shaped to fit the context of the prevailing cultural milieu. This is why they all look so superficially different. But cut away that social and cultural veneer and, underneath, sometimes very deeply hidden away, the same nucleus is often revealed, the same original, mystical message. All is one. All is Wholeness. And ultimately, the authentic goal of religion is to lead the individual into an experience of that Wholeness. This is not merely the wholeness pointed to by the new physics. This is an altogether more profound kind of wholeness—the spiritual seamlessness of our deepest subjective reality. When separated from their tyrannical dogma and metaphysical embellishments, all the great world religions seem to embrace the idea of an immanent God, a God that is to be found within rather than without. Aldous Huxley, following Leibniz, called it the 'Perennial Philosophy': 'the psychology that finds in the soul something similar to, or even identical with, divine Reality'.[3]

The Perennial Philosophy is not a belief in which we must simply, unquestioningly place our faith. It is not a piece of religious dogma. It holds no promises, no possibility of salvation, no reassurances to help cope with the suffering of life. It is an insight into the nature of reality which, when discovered experientially, has proved, throughout history, to transform the consciousness of the individual. The experience of Wholeness is a moment of enlightenment. It is a liberation. It is a discovery of a very special kind of freedom, a realization of a new, more profound sense of selfhood, a sense that transcends the individual ego.

We seem to have to pay a high price for our self-awareness. We seem to enter the world desperately alone, as if cast up on a desert island of consciousness, seemingly cut off from the rest of the universe, estranged, isolated, fragmented from the whole. Simply speaking, what all religions tell us, in some way or other, is that our existential solitude is only *apparent*. In the West, the function of popular religion has always been to act as a kind of bridge between our individual islands of consciousness and some 'higher' spiritual order. But the Perennial Philosophy represents a much deeper religious idea. It says that our estrangement is an illusion. Indeed throughout history, all kinds of mystical religious traditions, some operating from within the mainstream faiths, have had as their goal the unveiling of this illusion—to reveal the ultimate religious truth that all our individual islands of consciousness are joined as one. Instead of trying to build bridges, mysticism attempts to reveal, through direct experience, that we have no need for them.

There is a classic metaphor. Each one of us, as sentient beings, is like a window of consciousness in the vastness of the fabric of reality, opening into existence at birth, gradually growing as we mature into adulthood, and then, eventually, diminishing and weakly blinking out of existence at our death. The trouble is that we generally understand the metaphor the wrong way. We are taught to associate ourselves with the physical body of the window when we should be identifying with the 'light' that pours through it. Although all our windows are shaped and sized in an individually unique way, and are always being dulled to some extent by the opacity of ego–centredness—resulting in each having its own fluctuating signature of clarity—we are none the less all open to one and the same source of spiritual illumination. We are each the light, the *same* light. Ultimately, our individual selves are one Self. This is the Perennial Philosophy. Unfortunately, it cannot be embraced by rational thought. It can only be *felt*, intuitively. In the words of Alan Watts:

> It is as if the eyes were trying to look at themselves directly, or as if one were trying to describe the color of a mirror in terms of colors reflected in the mirror. Just as sight is something more than all things seen, the foundation or 'ground' of our existence and our awareness cannot be

understood in terms of things that are known. We are forced, therefore, to speak of it through myth—that is, through special metaphors, analogies, and images which say what it is *like* as distinct from what it *is*.[4]

And he continues:

In using myth one must take care not to confuse image with fact, which would be like climbing up the signpost instead of following the road.[5]

That is a wonderful metaphor for the state of most of our traditional religions. Having climbed the signpost instead of travelling the road, they are found to be sitting in a very embarrassing position, a position so embarrassing, indeed, that it takes an enormous amount of courage to come down and set foot back on the road. That is why, perhaps, the faithful still so doggedly believe in the literal truth of their religious myths—indeed, why they believe with such proselytizing fervour. Cravenly and precariously perched on top of the signpost, they seek ways to reinforce the immuring walls of their self-delusion rather than escape from it. Traditionally, the commonest method of bolstering faith in this way is conversion. The more people who believe in a particular religious dogma, the more 'true' it becomes in the eyes of its followers, and the easier it becomes to overlook its metaphysical contradictions. Ultimately, it is only by winning more and more 'believers' to its particular brand of faith that a religious system can endorse that faith. The ludicrous apotheosis can be witnessed in many parts of the undeveloped world where Christian missionaries of all kinds of persuasion are to be found embroiled in ideological battle with each other, zealously competing for converts among the native people—people who were previously very well served, in their particular cultural context, by their own far more meaningful spirituality.

Most of the great organized religions of the world contain a mixture of two essential kinds of myth: tribal and archetypal. Often the same myth can be both tribal and archetypal, depending on its interpretation, literal or mythic. Whereas archetypal myths refer to universal features of human experience, tribal myths refer specifically to just one particular 'tribe' of people, just one particular section of humanity. Whereas archetypal myths integrate people of different cultures by revealing the commonality of human experience, tribal myths have frequently served to discriminate, differentiating between people, demarcating one ideological group from another, and in the worst case functioning as a form of spiritual apartheid, granting a particular tribe the status of a chosen élite, with special rights and privileges. By adopting such a myth as a tenet of faith, a tribe can create a clear identity for itself, and can also define a second tribe, of unbelievers, upon which they can unconsciously project their fears and insecurities. Where archetypal myth seeks to break down the defences of the ego, tribal myth works by reinforcing them, and that is why the tribal interpretation of myth has always dominated the archetypal. It is a fundamental human weakness that we will never face a difficult truth unless we have to. The tribal interpretation offers an easy 'truth'—because it is really a fiction which preys on the vulnerability of the ego.

By contrast, archetypal myths help us come to terms with our existential fragility. In their symbolic interpretation, they can lead us into a confrontation with the very difficult truth represented by the full reality of the human condition, into an acceptance of our humanity in all its aspects and, ultimately, into an acceptance of the *mystery* of who we are. In moments of crisis, stuck in the world of ego, trapped in a self-woven web of anxiety, it is perfectly natural to feel anger at being brought into existence without our consent, dumped into being without a map, without a torch, without any idea of who or where we are. This is a very real and genuine experience. It is just not valid to ease it with the simplistic platitudes of traditional religion, for that only removes the anxiety from a conscious to an unconscious level where it can wreak all sorts of underground psychic havoc. To pick up again with Alan Watts:

> Irrevocable commitment to any religion is not only intellectual suicide; it is positive unfaith because it closes the mind

to any new vision of the world. Faith is, above all, openness—an act of trust in the unknown.[6]

Of course, faced with a world which offers no secure or tangible kind of value, such an act of trust in the unknown demands considerable courage. It is far easier, and far more comforting, simply to accept some pre-packaged religious idea about the way things are. By adopting a well-defined metaphysical map, such as the Bible or the Koran for example, which ostensibly offers a set of unimpeachable values, the faithful are relieved of the trying task of thinking and making decisions for themselves. This is why such a closed-off faith is indeed intellectual suicide. It amounts to a cowardly rejection of personal responsibility. The only authentic response is to open our mind and trust to the mystery of who we are, to accept and come to terms with the existential anxiety that is intrinsic to our humanity. In fact such an experience, a coming into contact with the reality of the tragic aspect of our being, is an essential prerequisite for a full, vital and authentic life. As so beautifully expressed by Kahlil Gibran in *The Prophet*, 'The deeper that sorrow carves into your being, the more joy you can contain.'[7]

Yet we need help in order to be able to assimilate these more difficult realities. Without some kind of guiding compass, existential despair can lead us ever deeper into a dark labyrinth of fear and insecurity. It seems that myth is the only such guide that we can trust with any kind of objectivity. By opening ourselves to its eternal wisdom, we can gain access to a kind of inner compass. According to Joseph Campbell, myth can put our conscious minds in touch with our secret, motivating, energizing depths.[8]

The images of myth are reflections of the spiritual potentialities of every one of us. Through contemplating these, we evoke their powers in our own lives.[9]

Myths are not truths. Myths do not have absolute meanings. They are magical mirrors in which we can look at the reflection of the meanings that we each carry with us in our soul. A myth is a very special kind of story. Whereas an ordinary story grows stale the more often it is told, a mythic story retains its originality—

because it speaks to our own originality. Indeed, no matter how many times we might listen to a great myth, we can never hope to get to the bottom of what it is saying. There is always more to be offered because we are not so much listening to a story but to the echo of our own intuition, the resonant voice of our innermost being. By steering us into a confrontation with the unknown and the uncertain, myth opens our mind to wonder. Myth offers a vision with which we can guide ourselves out of the labyrinth of self-absorption. By resonating with something deep within the psyche, myths allow us a glimpse of the map which encodes our human potentiality.

> They are telling us in picture language of powers of the psyche to be recognized and integrated in our lives, powers that have been common to the human spirit forever, and represent that wisdom of the species by which man has weathered the millenniums. Thus they have not been, and can never be, displaced by the findings of science, which relate rather to the outside world than to the depths that we enter in sleep. Through a dialogue conducted with these inward forces through our dreams and through a study of myths, we can learn to know and come to terms with the greater horizon of our own deeper and wiser self. And analogously, the society that cherishes and keeps its myths alive will be nourished from the soundest, richest strata of the human spirit.[10]

Our archetypal mythology is a collection of stories which through time have lost all vestige of original authorship and have therefore become timeless, surviving generation after generation simply because of the irrepressibility of their wisdom, a wisdom which is independent of race or culture or epoch. It is in the remarkable universality of these stories that we find, I believe, a guarantee of their validity as a guide and an inspiration in our lives. There is an inner consistency to the various streams of mythic thought throughout the world, a core of agreement, that a scientist would recognize as evidence of a deeply rooted intrinsic pattern. It surely has to be more than mere coincidence that the

Perennial Philosophy has appeared on metaphysical maps all over the world. What is clearly suggested is that all the great world faiths and mythologies are in contact with one and the same ultimate reality, all struggling to give objective form to the same inherently subjective message, namely that meaning in life is to be identified with fulfilling our natural human potential, with finding our own spirituality—within ourself. Their common faith is a faith in human nature, a faith in the human ability to reach within and find an immanent spirituality, a fountainhead of unconditional love.

Love has become such a heavily loaded word in our culture that we must be very precise here in saying exactly what we mean by it. In *The Art of Loving*, Erich Fromm proposes a definition which models perfectly the love that is expressed so magnificently in our mythological heritage. Fromm suggests that love always implies certain basic elements: care, responsibility, respect and knowledge. Care is self-explanatory and needs no clarification. Responsibilty is understood here in its original sense of feeling the need and being ready to *respond*. Respect and knowledge—or understanding, as I prefer—are essential in providing the context for care and responsibility. As a quality of love, respect is the inner strength to allow to be; it represents the absence of pity and possession and exploitation. Understanding—which necessarily includes self-understanding— provides the overall context in which care, responsibility and respect find meaning. Without respect, love is pitiful, possessive and exploitative; without understanding, love is blind and un- directed. *Love is the conjunction of care and responsibility guided by respect and understanding.*

Defined in this way, love is an art and a practice which requires a great commitment, and it beautifully captures the essence of the religious attitude to life. Spirituality is all about love. A spiritual person is someone who is in love with life, and who is seeking to deepen that experience, someone who cares, who feels a sense of responsibility, the need to respond—not for any reason that can be rationalized or dogmatically defined, but simply because there somehow exists an irresistible imperative in their psyche making it impossible not to care and want to respond. This kind of authentic love is intimately related to meaning. In fact, I would like to suggest that *love and meaning are the objective and subjective poles of the same basic experience*.

Love and meaning are complementary. Love flows naturally whenever we feel a sense of meaning; we feel a sense of meaning in the act of loving. Love is a natural consequence of being at 'one' with the world, of belonging. We naturally care about and feel a responsibility for our actions when we are able to identify with what we are doing, when we are in contact with meaning. Love and meaning feed upon one another, although unfortunately not always positively. In modern society, I fear, the feedback is often negative: when our sense of meaning is diminished, so is our capacity to love, which further diminishes our sense of meaning, and so on. This, very basically, is the root cause of our contemporary spiritual crisis. What we most lack today is a framework in which to understand and interpret our spiritual, existential experience. We cannot live authentically without some kind of life-support, the kind of support that is offered by myth. We need symbols on which to hang our innate spirituality. Our need for a religion is really a need for a spiritual mythology, to complement our scientific mythology, to take us beyond objective and subjective reality, to provide a deeper rationality through which to enter into the wonder of the bottomless mystery of existence.

The problem is that the kind of religion we are talking about here does not lend itself at all well to being organized in a systematic way. Nevertheless, throughout this century, there has in our society slowly grown up a religious sub-culture populated by people who have worked out their own forms of spirituality. Their numbers are probably far greater than we would suspect—and may well include many who still live within the loose embrace of Christianity. It is impossible to know for sure because they have no public voice or organization. They all simply share, although each in a unique way, the same inner sense of the sacred. The people belonging to this rising sub-culture have each awakened to the religious life, but, like myself, have found themselves estranged from the evangelism which tends to dominate popular perception of the Christian religion. As children of the technological age, the act of faith required to accept the dogma of Christianity has become just too great. It is seen to represent an unacceptable leap of irrationality, a leap no reasonable God could possibly expect us to make. Finding the traditional religions mostly irrelevant and inappropriate to our spiritual needs in a science-based culture, we have turned inwards to find a faith within our humanity.

Yet ours is still ostensibly a Christian society, with a tribal myth, a believing society where fewer and fewer people believe. Swelling undercurrents of unbelief are backing up against overcurrents of belief, causing, among the young especially, a great deal of epistemological bemusement. I would suggest that it is just this built-in social contradiction that has made religion such a taboo subject among so many people today. Religion touches the very nerve-ends of people's self-delusion. And it often hurts. The great problem we have in the West is that religion is invariably conceived in terms of God. Indeed, the two words 'God' and 'religion' have become virtually inseparable in our language. Anyone brought up in a traditional way will have been imprinted with an image of God as some kind of austere, grey-haired, berobed father-figure, an inerasable legacy of the nursery stories of early childhood. Like the billiard ball model of the atom, once implanted, it is a devilishly difficult image to banish from the mind. What picture does God immediately evoke for you? Do you find that you have to fight through this naive conception before reaching your more mature understanding? I certainly do, even now. In fact, the word 'God' has acquired such a vast complexity of connotations over the years that it has now become quite impossible to use it without evoking all sorts of spurious meanings and false images. 'God' has actually become a linguistic obstacle which is preventing us, as a society, from embracing a more profound understanding of religion.

The taboo surrounding death is another such barrier. The raw nerve-ends of our self-delusion can be very sensitive to the thought of our mortality, and for many traditional religion has always functioned as a kind of spiritual pain-killer, serving as a palliative by purporting to promise a life after death as reward for keeping faith during what amounts to a probationary apprenticeship on Earth. In this way, it often provides vital hope for people whose circumstances, personal, social or political, have denied them any other kind of hope, and that service should not be belittled, but it is all so very uncomfortable. A counterfeit sense of meaning is better than no meaning at all, but, at the same time, it arrests the development of any authentic sense of meaning. It encourages a perception of life as an objective possession rather than a subjective process, and therefore discourages people from working to improve the quality of their consciousness in this present life, and realizing its potential as completely as possible.

What most distinguishes humanity from other species on planet Earth is not so much our self-awareness—our being aware of being aware—but our being aware that unawareness awaits in the future, an unawareness of which we will be blissfully, obliviously unaware. Death itself can hold no possible fear for us. It should play a positive role in our lives, the complement to our birth and the most solid and absolute fact of our existence, an unavoidable destiny which gives life an essential sense of urgency. We each have but just the one short life, so we had better make the most of it. Time is precious. To accept this brutal fact of life is hard, but absolutely essential for the firing of our creative desire. It is just because we cannot indefinitely put it off until tomorrow that we are driven to create meaning now, in our relationship to spouse, family, society, culture—through *love*. Interpreted tribally, most religions promote a flight from life, and therefore a flight from the fears that life holds. Because our archetypal myths were originally seeded in those same fundamental fears, they can help us to confront them and to look into their true nature, therefore enabling us to discriminate between the genuine fears that we all share and the fake fears that we invent for our ego's protection. For only by allowing ourselves to be free to fear the genuinely fearful things of life can we let go of the petty fears that stop us living to our full potential. Suppressed, our fears can be suffocating. Accepted, they can release our natural creativity.

Working at a deep, unconscious level, it is our awareness of our impending unawareness that drives our desire to create meaning. Death and desire form an inseparable relationship with each other. And it is our attitude towards death—which cannot be divorced from our attitude toward life—that determines whether our desire is constructive or destructive. Authentically, freely flowing, desire engenders love. Blocked by self-delusion, it usually finds an outlet through power-seeking of some kind. The concept of a personal life after death is a spiritual crutch, and for most people an unnecessary, retarding, encumbering impediment to living a more authentic existence. Religion has surely to be wholly concerned with our present life. We must affirm the realities of the human condition, never deny them. And this, I believe, was very much the original teaching of Jesus. Jesus was, by living example, showing us that the path to spiritual integrity is trodden in *this* life, by a process of reconciliation and inner growth which leads to an outward manifestation of immediate, universal brotherly love.

One of the few things that does seem clear about the historical Jesus is that he was a truly great poet. Unfortunately, though, it appears that much of his most telling poetry has been lost to the world at large because it was censored during the compilation of the Bible. Many of the stories and sayings attributed to Jesus—like those from the Gospel of Thomas, one of many Gnostic texts which were discovered at Nag Hammadi in the Egyptian desert in 1945— were edited out of the Bible because they were too controversial. They contradicted the tribal myth that was being promoted by the political institution of the Christian Church. The chief aim then was to construct a universal religion, a *catholic* religion, open to all, a religion anyone could join provided they followed its doctrine and ritual, and recognized the clerical hierarchy as the divinely sanctioned carriers of God's message. Clearly, the Bible could not be seen to include material which suggested that the spiritual road was to be trodden subjectively. In fact, it was principally to silence the heretics, predominantly the Gnostic Christians who were proclaiming this subjective route as the true direction of Christianity, that the ideology of what we now know as the Christian religion was actually consolidated into the Bible in the first place.

By revealing religion to be essentially subjective, a process of transformation with no immediate tangible reward for all one's effort save a great deal of inner turmoil, the Gnostic interpretation of the teaching of Jesus did not represent ideal material for the foundation of a universal faith. The ecclesiastical Christians had to suppress the Gnostic heretics and their writings, and, of course, they succeeded almost perfectly, chiefly due to their very effective method: making the price of heresy the highest possible—death to anyone who refused to rescind. In simple terms, the Christian religion was designed to be easy to follow and self-validating, so that it could be open to all, and defended against all, which is still very much the situation as it stands now, although thankfully heretics are no longer burnt at the stake.

The life of Christ is a beautiful and profoundly compelling symbol, but can only have true meaning for us if we can relate to Jesus the man as opposed to all the supernatural luggage that has been pushed onto him and has become his lot to carry through his mythic life. If the story of Jesus is interpreted from a wholly human perspective it comes alive with a truly archetypal power. Through no other personal biography can we learn so much about our humanity, about love, fear, forgiveness, the spiritual life, and ultimately sacrifice. When interpreted archetypally, Christianity is perhaps the richest and most approachable religion ever to have evolved in human culture. For the genuine Christian, Jesus symbolizes an ideal, the human embodiment of a set of values towards which to strive, to be realized in our own life. The humanity of Jesus is experienced to understand our own humanity and its potential. His life, finally, is tragic, as finally is all human life. By compassionately entering into the suffering of Jesus on the cross, we are more fully able to come to terms with the unavoidable reality of the suffering that is inherent in all life, and are therefore able to open ourselves more fully to the authentic experience of joy.

To be religious is to be intoxicated with the *mystery* of life. It is to regard life as a question that needs answering. It is a seeking, a seeking after reconciliation, at-one-ment, with our existential situation. And that is all about ridding ourselves of our delusions, seeing through the simple answers that we accepted as children and searching beyond. I look on spirituality, not as a faith in God, but as a faith in human potential and human goodness, a faith in the validity of the meaning we create for ourselves through our love, a meaning which provides a shift of identification, from the individual ego to something far greater, far richer and far more profound.

Religion has ultimately to be personal and existential. It cannot be separated from the day-to-day business of living in *this* world, *this* universe. It must offer a path by which we can transcend our individual sense of nullity, not by promise of a better world in a life after death, but by helping us to achieve a state of mind at one with life and nature, a state of mind able to revel in the fantastic, inescapable mystery of existence, able to glory in its wonders and accept its inevitable tragedies. It must be all about helping us to face the reality of our existence instead of escaping it. In other words, it must be all about encouraging our spiritual awareness, and it is this awareness which triggers that positive feedback loop between love and meaning. As we become spiritually aware, we find ourselves accepting the responsibility inherent in the bit-part we each play in the vast performance of reality, recognizing the vitality of our assignment in the developing process that is the universe. Aware of the irreducible patterns which connect each of us into reality, both inwardly and outwardly, we experience meaning through understanding that our thoughts and actions will be forever imprinted in the future path that this process follows. This is our immortality—our *only* immortality.

This definition of meaning is perfectly rational and can happily be taken aboard by most people without any major compromise of personal metaphysical belief. We have already seen how science, through the new physics, has endorsed the mystical vision of the universe as one unitary process. Yet still, perhaps, the heart yearns for something deeper, some sacred sense of meaning. And I believe that there is such a sense of meaning to be understood, not through the intellect but through that more deeply rooted part of our being, our intuition. I am speaking here of an understanding of meaning that totally defies any kind of rationalization. In this pre-intellectual sense, meaning is experienced through feeling, through a feeling of identity, the identity of our being with the 'being', the Mind, of the universe—the feeling that, in some sublimely ineffable way, we *are* the Mind of the universe. This is the intuitive, numinous experience which is the Perennial Philosophy.

It was some such existential sense of the sacred to which I was awakened in Yosemite, in the encounter with the little humming-bird described in the Prologue. I seemed, that day, to make connection with a form of knowledge that was previously unknown to me. I became aware of being linked into a pattern which connects at a level far deeper than that to which rational analysis will ever be able to penetrate. This is why I could only look to mythology in order to interpret my experience, through that vast reservoir of wisdom which reflects the Perennial Philosophy in the experience of different human cultures throughout the world and throughout history. An epistemological whirlwind had blown through me, transforming the context of my life, throwing up far more questions than answers, leaving my old belief system in tatters. Myth helped me to rebuild by providing a blueprint for the construction of a new framework of belief, a new personal mythology.

To borrow a beautiful expression from Joseph Campbell, we are the ears and eyes and the mind of the universe, yet how often do we truly live through our ears and our eyes, and our feelings? How often are we truly aware of what we hear and see and feel? Myth seeks to turn our attention away from the churn of our preoccu-pying, mundane, everyday thoughts, back in touch with that primitive, archetypal desire for meaning—to hear and see and feel in a pre-intellectual sense, originally. It seeks to awaken us to that greater part of our existence to which we are normally asleep. From childhood through adolescence and into adulthood, myth can be our guide and our mentor, our staff and our compass, on the journey of self-confrontation and self-discovery that is spiritual growth. The immutable wisdom of myth, shocking, supporting, steering, seeks to disperse the clouds of our delusion. Out of space, out of time, myth is free of the socially created prejudices which have invariably corrupted the great religious movements of the world. Before myth, every man and every woman on Earth stand on the same footing, equally franchised to find the sacred within their own heart. The sacred is not to be found without. It is the divine spark which flares in us all. This is the inspiration of the authentic mythology of religion.

5

The *Tao* of Wisdom

ACCORDING to an ancient saying of the East, 'when the pupil is ready, the teacher will come'. I became ready, I believe, after that experience in the Yosemite forest where my consciousness seemed suddenly to awake from a long hibernation. I experienced a movement from within, an openness to self-exploration, a desire to grow. 'Each of us guards a gate of change that can only be unlocked from the inside,'[1] suggests Marilyn Ferguson. And that is just what happened following my encounter with the humming-bird: the defences of reason came down and allowed my intuition to sneak in and unlock my 'gate of change'. I became ready to learn. The teacher that came to me was the *Tao Te Ching*.

Written about 2,500 years ago by a Chinese sage whom we have come to know as Lao Tzu, quotations from the *Tao Te Ching* started bombarding me during the period of rapacious reading that followed my return from California. Serendipity was telling me that this was a book I had to read. I soon found, though, that this mythic teaching was going to offer no magical solutions. In fact, its strange, cryptic structure seemed at first to be designed to be deliberately obfuscating. To a mind that had been reared on traditional Aristotelian logic, the adamantine paradoxicality of the *Tao Te Ching* was rather bewildering, like being introduced to the music of Messiaen after having only ever heard Mozart. The breakthrough came when I stopped trying to intellectualize its content. Only then did I begin to find my way in its impenetrability and come to appreciate its very subtle, usually implicit meanings. By stilling my restless, analytical mind, I allowed my intuition to engage with its haunting wisdom. Here is the very first and very

mysterious chapter, taken from the brilliant contemporary translation by Stephen Mitchell:

> The Tao that can be told
> is not the eternal Tao.
> The name that can be named
> is not the eternal name.
>
> The unnameable is the eternally real.
> Naming is the origin
> of all particular things.
>
> Free from desire, you realize the mystery.
> Caught in desire, you see only the manifestations.
>
> Yet mystery and manifestations
> arise from the same source.
> This source is called darkness.
>
> Darkness within darkness.
> The gateway to all understanding.[2]

One of the great fascinations of the *Tao Te Ching* lies in the
intricate problem of its translation into modern language. Meanings
balance so delicately upon the subtle interplay of words and phrases
that they cannot help but become tilted in translation, especially
when one takes into account the peculiar idiosyncracies of Chinese
calligraphy and the fact that the original characters have changed
and evolved throughout history, not just in shape and form but also
in use and meaning. An objective translation of the *Tao Te Ching*
is unapproachable. Every version has a distinct character stamped
on it by the particular nuances of the translator; in fact, it is this
subjectiveness, the inevitably frayed edges of our modern representations, which makes the work still very much alive today,
despite the passing of two whole millennia.

The only way to reach back to Lao Tzu is by reading as many
translations of his work as possible. Out of a wide variety of
different interpretations and word selections, a meaning often
emerges which could never be encapsulated by a single label or
phrase. Reading the *Tao Te Ching* in this way, it soon becomes

clear that English, in spite of its incredible range of linguistic sources, is sadly lacking in its depth of spiritual language. Also, since the spiritual language we do have is so irrevocably tied to the Christian religious tradition, the English vocabulary of the sacred inevitably carries associations which tend to alienate the atheistic mind. In fact, the great appeal of the *Tao Te Ching* to the modern Western mind may well have much to do with the freedom of its key concept from traditional connotations. The *Tao*—pronounced 'dow' as in 'dowel'—is consumed by an intrinsic mysteriousness, transcending all language, beyond translation, inscrutable, ineffable. Any attempt to describe it in rational terms is doomed to failure.

The *Tao* is the deepest and most fundamental of all nature's patterns. It is the underlying evolutionary principle of the universe. The ultimate impulse to creation, prior to object and subject, prior to all dichotomizing categories.

Look, it cannot be seen—it is beyond form.
Listen, it cannot be heard—it is beyond sound.
Grasp, it cannot be held—it is intangible.

From above it is not bright;
From below it is not dark:
An unbroken thread beyond description.
It returns to nothingness.

The form of the formless,
The image of the imageless,
It is called indefinable and beyond imagination.

Stand before it and there is no beginning.
Follow it and there is no end.
Stay with the ancient Tao,
Move with the present.

Knowing the ancient beginning is the essence of Tao.[3]

Taoism, the spiritual stream of thought which sprang out of the *Tao Te Ching*, does not really represent a religion in the sense that word is commonly understood in the West. It has no systematic teaching, no creed, no moral code, and certainly no dogma. In fact, the *Tao* has an alternative meaning in Chinese which translates as 'The Way'. Taoism represents a way of life, and can only be understood in that context—through being lived. It is based on experience rather than faith. There is no imposition from without, only a striving to be true to our own inner nature, to conform to the harmony of the *Tao*. Taoism steers us towards a discovery of who we are. It challenges us to use our *own* awareness of reality, our *own* wisdom, rather than relying on other people's second-hand opinions. It challenges each of us to find our own way through life, to discover our relationship to the universe, free of any fixed ideology.

This is what, for me, makes Taoist thought so exciting. It is all about achieving a *right* state of mind, and consequently a way of *right* action and living, in the present moment. It aims at achieving strength and independence of mind by provoking us, often through paradox, into a direct confrontation with the reality behind the appearances. It teaches us to understand and accept that the only place we ever exist is at the point of flux where the future meets the past. It is in this perpetually transforming moment of time that the dynamic play of reality is enacted. We are to learn that our part in this movement must also be dynamic, ever responding, ever adapting, never resisting. We are to learn to accept life's essential impermanence, realizing that death is part of a natural cycle which makes our existence complete. Life requires death for its completion. Life and death are complementary. Taoism teaches us the folly of seeking security in permanence. True security is only to be found in the freedom achieved through forsaking attachment. In Taoist philosophy, the world of appearances contains no absolutes. There is no good and evil. Sin is seen to be simply ignorance, a lack of understanding of our own inner nature and its identity with the *Tao*. For it must surely be pure stupidity to act deliberately out of harmony with the universe because that act is inevitably reflected back to us in terms of inner conflict. To injure the natural harmony of the *Tao* is simply to injure ourself. To violate the natural harmony of the universe is simply to violate our own humanity.

At the heart of Taoist philosophy lie the simultaneously complementary and opposing concepts of the yin and the yang. Indeed, the principle represented by their mutual interaction is of far greater antiquity than the *Tao Te Ching*, preceding it by over 2,000 years. It is deeply fundamental to all Chinese thought and culture. Yin and yang do not represent an absolute dualism, a common feature of many Western systems of thought, but a relative dualism, an utterly dynamic interplay of cyclic poles, reciprocal but complementary, in tension but also in harmony. Their interaction is the ceaselessly shifting, perpetual flux that is the *Tao*, the creative rhythm of the universe. The two archetypal poles of yin and yang do not stand alone in absolute separation, but share the same common ground as opposite aspects of one whole, the one dissolving into the other in a process of cyclic change, the seed of yin contained in yang, the seed of yang contained in yin. This concept represents an ancient recognition of a message which arose very clearly from our study of the new physics—complementarity.

Despite the enormous temporal and cultural gulf, Taoist philosophy has just as much relevance for us in the West today as it did for the ancient Chinese, for the simple reason that the human condition and situation have not changed. Taoism strikes at the very crux of the malaise which has always infected civilized humanity. It is stated as the dissonance between mind and body, reason and intuition, the conscious and the unconscious, the conflict between the arid, delimiting nature of thought and the passionate, nebulous nature of feeling, the friction between the intellectual and the emotional, the objective and subjective facets of our being. The great existential problem which faces all humanity is seen in terms of the quest for inner harmony: attaining an equilibrium between the yin and yang aspects of our psychic nature. It represents the ultimate of balancing acts.

The inner accord of yin and yang is the *Te*—pronounced 'der' as in 'order'—of the *Tao Te Ching*. It is *Tao* manifest in human being. It is variously translated as virtue and power, but I much prefer integrity, a kind of inner strength of mind, the soft and supple strength which arises out of a paradoxical union of independence and vulnerability, the strength we feel when the inner pattern of our mind is in harmonic resonance with the outer pattern of our physical situation, when we are 'in tune' with what we are doing, when we are 'at one' with our actions, when things are flowing, confidently, effortlessly, without the interference or inhibition of the ego. *Te* is the integrity of a mind unimpaired by self-consciousness, unclouded by self-delusion. In Taoist literature, this psychic balance of yin and yang is found embodied in the Sage or Master. The Taoist Master is the consummate example of *Te*, the Wise One, strong and gentle, receptive and adaptable, dignified and humble, the ideal individual, living in yin-yang harmony, head working with heart, intellect with instinct, analytical thought balanced by gut feeling, knowledge balanced by value. This integrity of being is the spiritual goal of Taoism—wholeness, holiness, healthiness, the same root signifying the same meaning, the same ideal.

In addition to its mystical content, then, the *Tao Te Ching* also offers a practical guide to living a more authentic existence. At the centre of this philosophy of life is the concept of *wu-wei*, the principle which connects *Tao* with *Te* in the individual person. It represents a yielding of the ego, a 'letting be', an avoidance of action that is contrary to the natural way of the *Tao*, a respect for the natural order, the natural patterns and rhythms of the universe. Forceful action, imposed against the natural movement of the *Tao*, inevitably creates its own forceful resistance. Action meets reaction. The passive action of *wu-wei*, however, seeks to harmonize with the rhythm of nature and therefore meets no opposition. In relationships, our actions should always be guided by a respect for the *Tao* in the other person, allowing them to manifest their own *Te*. When a couple let each other be to follow the *Tao* within themselves, they enter into a dance of the spirit, bringing each other more fully alive. This is the Taoist way of life, the path towards wholeness. According to Lao Tzu, as translated by Stephen Mitchell, 'Only in being lived by the *Tao* can you be truly yourself.'[4]

My initial problems with the *Tao Te Ching* were mostly caused by my deeply rooted desire to rationalize its concepts. I wanted to picture the *Tao*. I wanted to systematize it, contain it within a box of reason. Having come out of a kind of existential sleep, I was overcome by a naive and almost avaricious hunger for new understanding. I wanted answers to my existential questions. I wanted to resolve paradox. In time, though, the *Tao Te Ching* helped me to confront the impossibility of these early ambitions. It helped me to reach a level of maturity where I could accept the infinite dimensions of paradox and live at some kind of peace within its domain. As Lao Tzu says, 'The more you know, the less you understand.'[5] To attempt to contain Taoist philosophy within a metaphysical system of explanation is to kill it. It is also to fail.

The *Tao* is to be experienced, not talked about. Yet the concept is so foreign to our cultural tradition that some amount of talking is perhaps necessary to ease the way a little, as long as it is remembered that any words or pictures, because they are unavoidably haunted by that ghost called reason, can only provide us with a very crude intellectual handle with which to facilitate the much more subtle grasp of our intuition. My basic conceptualization of the *Tao* is as the *will to being*. In terms of the classic metaphor introduced in the last chapter, it is that vital, animating, spiritual 'light' that seeks to radiate through the windows of awareness that we each represent in the physical fabric of reality. For me, one of the principal messages of Taoism is that the insecure, maldeveloped ego functions like a smoke screen on this window. When we get caught up in our own selfish concerns, the glass clouds over. Trapped by petty fears and anxieties, we lose our radiance. We lose our ability to love and create meaning. We lose our capacity both to extract joy from life and to engender it in others. Only by relaxing the hold of the rational mind, so giving mental space to our intuitive mind, can we clear the fogginess and let some light into our gloomy prison of self-absorption.

Since before time and space were,
the Tao *is*.
It is beyond *is* and *is not*.
How do I know this is true?
I look inside myself and see.[6]

After reading the *Tao Te Ching*, I always get the feeling that I have been put in touch with some very profound wisdoms of the human condition. I know what kind of person I should be striving to be. I know where I should be going on my journey of growth. But I'm never left feeling sure of just how I go about getting there, how I should become this ideal individual. The theory always sounds good, but as soon as I leave the world of ideas to return to the world where the mortgage has to be paid, the practice seems to be impossible. Today's spiritually shrivelling environment is simply not conducive to the way of life advocated by Lao Tzu. It is all very well adopting a philosophy of 'let be', but when everyone else is pushing and shoving there seems to be a very real danger of being trampled upon in the mad rush. And Taoism, for me, doesn't seem to have anything useful to say about the violence and suffering that inheres naturally within the rhythms of life. Do I save the little bird that is being teased and tormented to death by the cat?

What has to be understood is that the mythology of Taoism is not a complete answer to the spiritual questions of life. No single mythology can fulfil that role. All we can expect from any mythology is a partial understanding of our humanity, an insight into just some of its countless aspects. We have to be eclectic in our seeking. What I most picked up from the *Tao Te Ching* was a respect for my own individuality. For a while, I wanted to label myself a Taoist, but when I came to appreciate the wider meaning of Lao Tzu's philosophy and some of its contradictions, I slipped free of this desire. The *Tao Te Ching*, unlike almost any other work of philosophy, points very deliberately to its own inadequacy, and even its own irrelevance. We are our own teacher. We only require the courage to look inside our own being.

For me, the most important and practical insight offered by the *Tao Te Ching* is its view of the spiritual quest in terms of seeking a balance between the yin and yang aspects of our inner nature. Taoist thought recognizes the folly of developing one side of our nature at the expense of its complementary partner. Opposites have to be integrated into a dynamic whole. This view fits remarkably well with the model we built earlier (see Chapter 3). Following the discussion of Roger Sperry's investigation into split-brain patients, we viewed the psyche, in the light of modern scientific evidence, in terms of the two hemispheres of the brain, the left brain being home to quantitative, linear, rational perception, with a reductive, discriminating, ordering function, and the right brain being home to qualitative, non-linear, intuitive perception, with a deductive, synthesizing, patterning function. Very crudely, the left brain is home to our masculine yang nature and the right brain home to our feminine yin nature. The yang is the conscious mind, the 'mental' pole of our being; the yin is the unconscious, the 'physical' pole. It must be said that this model is a gross simplification of the true complexity of the human self, but it still offers, I believe, a valid framework in which to understand some of the most basic problems of human existence

In the moments following my encounter with the humming-bird in the Yosemite forest, my consciousness was flooded with meaning. Why is it that I couldn't hang on to the values that were so immediate to me then? Why is it that at other moments I am overcome with meaninglessness, where life seems pointless, devoid of value? Why is it that the vast majority of my life seems to be caught in suspension between these two extremes, where life does seem to have value and meaning, but not the urgent, vital, overwhelming, joyous kind of value and meaning that I know it should have, and potentially could have—if only I weren't too tired, or too afraid perhaps, to make the necessary effort? These particular questions have been the central stimulus to the work of the British writer and philosopher Colin Wilson. In *Beyond the Outsider*, he states our problem as follows:

> Man is in the position of a painter painting a gigantic canvas. If he is close enough to be able to work, he is too close to see it as a whole. If he stands back to see it as a whole, he is too far away to use his paintbrush.[7]

In fact, I believe there is a very fine point at which we *can* both be close enough to use our paintbrush and far enough away to see the overall effect. This is the state of mind in which the glass in our window is crystal clear, allowing the light of the *Tao* to stream through without diffraction. It is, however, an almost impossibly difficult point to centre upon, as if the smallest upset in circumstance or the slightest lapse of concentration will throw us off balance. Indeed, stepping back any kind of distance to see the greater picture of things seems to be very difficult. Another metaphor might be more appropriate here. Perhaps we should think of our window as being fitted with a Venetian blind, a blind which refuses to stay open on its own. The blind has to be held open by a deliberate act of will, an effort we tire of very quickly. It is far easier to relax and paint with our nose up against the canvas, and this, of course, is just where we spend the very great part of our life, daubing on paint in a careless, cavalier way, without any sense of context or overall meaning, disconnected from the whole. It is only when we meet some obstacle that we are forced to step back and gain a broader perspective. One of Colin Wilson's key concepts is what he calls the 'indifference threshold'.

> One of the most curious characteristics of human beings—particularly westerners—is that *pain and inconvenience stimulate their vitality far more than pleasure*. In a very precise sense of the word, human beings are spoilt. A spoilt child is one who has come to expect certain privileges and accepts them as rights. He is not grateful for these privileges; in fact, they bore him. The only time he feels strongly about them is when they are curtailed; then he sulks.[8]

There are many examples of people who have spent most of their life in retreat, seeking out boredom rather than joy, only finding some vitality and meaning to their life after a death-threatening or incapacitating illness. It often takes the threat of the loss of a freedom to make us appreciate and use it. We don't appreciate our capacities until they are taken away from us, and then we become so absorbed by resentment and regret that we fail to make use of the capacities that we still have left. Life is so precious, so extraordinary and so short, that we ought constantly to be living it to the full, but somehow, it doesn't seem to work out that way. The sense of urgency, the sense of the essential excitement of life, is never quite strong enough to motivate us as it should. The problem seems to be simply that we spend too much time too close to the canvas. We become stuck in our rational, analytical, yang-self, in a world denatured by thought. Because our yin-self has a yielding and responsive nature, it easily concedes to the louder voice of its dominant partner, innocently gullible to all the delusions which arise out of its narrow, limited outlook. In effect, the quality of the right-brain's wide-angle view is vulnerable to the shifting mood of the image which is seen through the microscope of the left-brain. If we are not vigilant, we can allow the clinical logic of our yang-self to put our yin-self to sleep, with the result that our perception becomes disconnected. Cut off from any contextual meaning, we can easily slip into a depressed state of consciousness, tired, moody, listless, bewildered by the lack of a sense of purpose. Instead of being turned out, reaching towards reality, our consciousness becomes turned in. We become bound up in ourself, our yang-self, the world of our ego, our thoughts continually churning over, continually turning the screw on a self-induced inner tension. We become bored. We lose our enthusiasm for life.

But what exactly do we mean when we say we've lost our enthusiasm? What is it that we have when life feels good that seems to desert us so completely when things are going against us? In fact, the word 'enthusiasm' has come down to us from the Greek *enthousiasmos*, which means, literally, 'filled with *theos*', or 'filled with God'.

I had never appreciated this rather wonderful etymology until I read Robert Pirsig's amazing *Zen and the Art of Motorcycle Maintenance*, a cult book which, unusually, fully deserves its cult status. It does not matter whether you love it or hate it, you cannot help but be stimulated by Pirsig's strange story. I loved it—in a hateful sort of way—and was certainly stimulated, especially by Pirsig's concept of 'gumption'.

A person filled with gumption doesn't sit around dissipating and stewing about things. He's at the front of the train of his own awareness, watching to see what's up the track and meeting it when it comes. That's gumption.[9]

The gumption-filling process occurs when one is quiet long enough to see and hear and feel the real universe, not just one's own stale opinions about it. But it's nothing exotic. That's why I like the word.[10]

And that's why I like the word too. We can all relate to gumption. Or enthusiasm. It's the inner strength, the *Te*, we feel when things are flowing, when we're running with the tide, when our yin- and yang-selves are on the same wavelength, resonating with each other. Pirsig sees gumption as a kind of psychic petrol. It is available to everyone, abundantly, yet, most of the time, the fuel lines seem to get blocked somewhere along the way. In our normal workaday lives, the gumption just doesn't seem to get through.

If we think about it honestly, our everyday consciousness is abysmally limited in performance. To use another motorcycle metaphor, it functions as if it has been taken out of gear, disconnected from its drive shaft. Here is Colin Wilson again:

> For human beings, boredom and depression are abnormal—
> a failure to grasp their natural powers. My powers are wasted
> so long as my vision is narrow and personal. They are like a
> boxer who cannot get any force behind his punches at close
> quarters. And when my will has become passive through
> 'close-upness', I fall into a dreamlike state in which illusion
> and reality are intermingled. I become trapped and tangled
> in my own narrow values, instead of remaining open to values
> that are greater than myself.[11]

Basically, what he is saying is that we are failing to live to our full potential because our gumption supply gets blocked. Again, our big problem is the short-sightedness of our yang-self. Its attention is easily diverted. It is continually being distracted by the trivial demands of the moment. And as it loses its way, so our yin-self responds with a yawn. It gets bored. The Venetian blinds close. The gumption supply dwindles. As Colin Wilson suggests, the real problem of human existence is the power of habit to rob us of our sense of reality. We too easily get trapped within the autonomous world of our yang-self, caught within the loops of its rigid programming. We get locked into habitual patterns of thought and behaviour. Our attention wanders. Our values slip out of focus. Our goals drift out of sight.

Wilson emphasizes the vital fact that consciousness should be intentional. Our ordinary perception is limited because we fail to put any effort behind it. Without intention, consciousness is unfocused. We become merely passive observers of our habitual responses. We behave as if we were nothing more than a system of conditioned reflexes. We become enslaved to our thoughts and emotions instead of being their master. And, tragically, we accept this impoverished, blurred, depressed state of consciousness as normal. It is almost as if we spend most of our time just sitting back and watching the play of the thoughts and events of our life on a television screen—and we have forgotten that we have the remote control in our hand. We can switch channels. We can adjust the emotional colour and contrast. We don't have to sit brain-numbed watching repeat after repeat of our own tedious soap operas. I am not suggesting that we deny our emotions, only that, once acknowledged, we let them go, that we don't needlessly hang on to them, that we don't wallow in them.

The trouble seems to lie with our autopilot. Its essential function is to take care of our day-to-day mental housekeeping; the danger is that in relieving us of the need to attend to the more repetitive, base-line activities of life, it also relieves us of the capacity to appreciate the vitality of the world around us. A child experiences joy and delight in simple things that we usually fail to admit into consciousness, because years ago the novelty was lost to us. The autopilot censors our awareness. It functions as a kind of filter on reality, admitting only that which it considers important. Think of the times you have glanced at your watch to check that you are not late for something, only to find yourself having to look again a few seconds later because you never noticed what the time actually was. It was only important to admit into consciousness the fact that you were not late; the exact time did not matter. As Colin Wilson points out, 'seeing' is not enough; it is necessary to mentally 'grasp' the content.

The autopilot has an essential role to play, but it must not be allowed to assume too much responsibility. When we are interested in what we are doing, when something is holding our attention, the autopilot stays in the background and our awareness is fully alert. The mind is reaching out and grasping content. Inevitably, though, our interest wanes. Boredom and tiredness wash over us. The autopilot takes over more and more control and, in Wilson's words, the reality around us becomes less and less real. As the autopilot assumes full command of consciousness, the outside world no longer exists for us. We cease to grasp. We become trapped in unreality. Nothing seems worth the effort any more.

> When I have to carry out some task, I 'summon' energy, and the right-brain obligingly provides it. While I am *deeply absorbed* in the task, there is almost no waste of energy. But if I begin to lose concentration, to get bored, it is as if the connection between a hosepipe and the garden tap began to work loose, so that half the water gets lost in spray around the tap.[12]

When we become absorbed in something, we are 'taken out of ourself', and that is in fact just what happens: we are taken out of our yang-self. We enter a mode of consciousness where our yin-self is playing an unusually active role. The energy it supplies is Pirsig's gumption. It is intimately related to our will. Our enthusiasm for life, our gumption level, is in direct proportion to the intensity of our will, the degree to which our Venetian blinds are pulled open. And it is stimulated by meaning. Meaning winds up our mental mainspring.

> The deeper my sense of the 'meaningfulness' of the world, the fiercer and more persistent my will. And increased effort of will leads in turn to increased sense of meaning. It is a chain reaction. So is the reverse, when 'discouragement' leads me to stop willing, and the passivity leads to a narrowed sense of meaning, and the gradual loss of 'meaning' leads to further relaxation of the will. The result is a kind of 'down staircase'

of apathy. On the other hand, any intense glimpse of meaning can cause a transfer to the 'up staircase'.[13]

This input of meaning can be quite ordinary. If we are doing any job of work that really matters to us, we seem able to summon up the energy with ease. The gumption is there. We are able to find the necessary patience and discipline. The will is tight. However, the same job, if it does not matter, if it holds no meaning for us, will drag because we cannot seem to summon the energy. The gumption is not there. The will is slack. It lacks a sense of purpose. We are at our best when we are living authentically, when we have some greater purpose to captivate and grip the will to a higher tension. We need the sense of meaning that comes with such a purpose just as an engine needs fuel. The will is fuelled by meaning. It is as if those Venetian blinds are sprung so that they are always trying to close against our will. Again, the problem seems to be the habitual, short-sighted nature of our yang-self. It requires a deliberate concentration of the will to increase the depth of focus, to see beyond the over-subjective view given to us by our 'normal' consciousness. That everyday view is highly capricious, and often spurious.

Wilson suggests that we live in a kind of room of subjective emotions and values. When we are trapped with our nose up against the canvas, the blinds close, shutting out the light that we need in order to be able to view our subjective values in a wider, objective, meaningful context. When we find ourselves succumbing to the fickle subjectivity of our emotions, or the socially conditioned responses of our reflex feelings, when the gate closes on wisdom, we need to be able to take a step back from the canvas. The big problem today is that very few people have learnt how to do this. The great majority of us live in an urban landscape with no rapport with the natural world, continually surrounded by the sophisticated apparatus of modern living, and there is very rarely any opportunity to stretch the sight toward a horizon. Day-to-day life is invariably acted out within the insulated confines of closed, human-made spaces: the house, the car, the office, the shopping mall. Sealed off from nature, many people's routine existence has become pervaded by an oppressive sense of 'close-upness', their sense of deeper purpose, their vision—the greatest of all human faculties—imprisoned within an artificial world of material comfort, surrendered to atrophy.

The prevailing value system of modern Western society is profoundly out of balance. Yang aspects rule over yin aspects. Masculine dominates feminine. Reason overshadows intuition. Forcefulness and aggression are rewarded far more than responsiveness and sensitivity. Material concerns engulf spiritual concerns. Status is gauged by quantity—salary and material wealth—rather than by quality—contentment and fulfilment. Lives are measured by outward success as opposed to inner well-being. The result is that society cannot help but impart this intrinsic dissonance into the psyche of every individual. Social imbalances inevitably filter through to cause imbalances at a personal, psychological level. They amplify the innate problems of existence, intensifying the felt sense of separateness and nullity. In order to cope, we are virtually compelled to protect ourselves by self-delusion, camouflaging our difficult existential questions, only allowing ourselves to see what we want to see, hiding our loss of identity with the eternal *Tao*, masking our deep yearning for an authentic sense of belonging. In this disconnected state of consciousness we lack any universal sense of meaning, making it possible for us to go to absurd, frightening lengths to defend our delusions. We disinherit our intuition. This is why, despite our clear understanding of the appalling social and environmental effects of the profligate way in which we live, we continue to do nothing to restore even a semblance of balance.

The vital importance of the main discussion of this chapter should now be clear. For without the combined gumption and wisdom of millions of individual people we will never have the strength, nor the vision, to heal the terrible wounds that we have inflicted upon the Earth. Susan Griffin speaks about this with an impassioned voice which never fails to stir me out of complacency.

We who are born into this civilization have inherited a habit of mind. We are divided against ourselves. We no longer feel ourselves to be a part of this earth. We regard our fellow creatures as enemies. And, very young, we even learn to disown a part of our own being. We come to believe that we do not know what we know. We grow used to ignoring the evidence of our own experience, what we hear or see, what we feel in our own bodies. We come into maturity keeping secrets. But we forget this secret knowledge and feel instead only a vague shame, a sense that perhaps we are not who we say we are. Yet we have learned well to pretend that what is true is not true. In some places the sky is perpetually grey, and the air filled with a putrid smell. Forests we loved as children disappear. The waters we once swam are forbidden to us now because they are poisoned. We remember there was a sweet taste to fruit, that there used to be more birds. But we do not read these perceptions as signs of our own peril. Long ago we gave up ourselves. Now, if we are dying by increments, we have ceased to be aware of this death. How can we know our own death if we do not know our own existence? We have traded our real existence, our real feelings for a delusion. Instead of fighting for our lives, we bend all our efforts to defend delusion. We deny all evidence at hand that this civilization which has shaped our minds is also destroying the earth.[14]

In order to know our own existence we need to be able to step back from the canvas. This is how we re-enfranchise our intuition and so recognize the evidence of our experience. This is how we open the gate on wisdom and so find a source of objectivity. We regularly need to put the paintbrush down for a while and temporarily withdraw from the hurly-burly of life. We need to find a meditation. The fact is that the voice of our intuition is not easy to pick out even when we are prepared to listen; if our ears are closed by habit of mind, it is virtually impossible to hear. The inner peace, the concentrated awareness—the Silence—of meditation gives us an invaluable interlude of respite in our hectic lives, a chance to recharge the batteries, to refresh the harassed soul. I have centred upon two main kinds of meditation in my life, two very different ways of seeking Silence. The first represents my way of harmonizing mind and spirit, and simply takes the form of retreating to a quiet place and reading, perhaps from one of my favourite mythologies, like the *Tao Te Ching*, or from the words of one of my spiritual heroes, people like Laurens van der Post or Joseph Campbell. Their wisdom never fails to supply a corrective balance to my self-perception when it gets distorted by the inner conflict born of fear and doubt. To feel the impetus of a mind which is in touch with the mythic patterns of human potentiality is a sometimes necessary catalyst to being able to touch that potentiality myself.

My second kind of meditation, as I said earlier, takes the physical form of walking or running, and represents my way of harmonizing mind and body. I like to get out in the countryside, most especially the hills or the mountains, or along the coast. Travelling alone and on foot through a wild landscape, I find that I can purge myself of worries and other calculating thoughts. The thinking mind is willingly sedated, the normally discordant voices of my inner world encouraged to harmonize with the insistent, anodynic rhythm of nature's very special Silence. There is a unique joy to being in wilderness country which is the joy of being taken, seemingly without effort, into this beautifully serene state of consciousness, body integrated with mind integrated with spirit. It is here that I always feel most deeply myself. Standing on top of a mountain peak or a sea cliff, the imagination is able to gain an escape from the daily exigencies of modern life. Released from the manacle of mundane necessity, it is free to stretch out, free to gaze into far greater possibilities than can ever normally be glimpsed while preoccupied with the humdrum grind of earning a living at the office or the factory. A wild landscape asserts itself upon my consciousness in a way that allows the imagination itself to run wild, to run primitive and naked, stripped of its social and cultural clothing.

The archetypal power of landscape, then, much like the power of myth, lies in its ability to provide the exercise that our vision so surely needs. It is able to uplift our spiritual perspective and entice the imagination out to play. At the focus of a mountain panorama, especially when the rational mind has been stilled by the rhythm and effort of running, I am somehow drawn into the landscape which contains me, aware of my part in the whole, as the origin of the enigmatic mirror in which its vast expanse is being so creatively captured. I feel bonded with the Earth. In the naturalness of the landscape, I see reflected my own naturalness. In the dignity of its untamed grandeur, I see reflected my own human dignity. As the imagination simultaneously reaches out and reaches in, wrestling with the paradox, there is sometimes a momentary sensation of seamlessness, of a sort of perfect continuity of the landscape with the mindscape, a fusion, a sense of oneness, wholeness, the *Tao*. And it is accompanied by an amazing feeling of being struck between the eyes with the utterly outrageous improbability of it all. The experience is wonderfully therapeutic, vivifying, allowing me to recover a sense of place and belonging. The problems of my everyday life contract to their true dimensions. I am able to grasp what is *really* important—the astonishing miracle of existence. I am left feeling immensely privileged to be alive, and to possess this incredible gift called imagination, ablaze with wonder. I am left determined not to squander my being in this world.

These meditations serve to grant me an experience of the reality of my being, and function as a marvellously restorative antidote to the terrible sense of unreality—that sense of 'close-upness'—that inevitably pervades so much of modern life. When I am trapped in that dark room of subjective emotions and values, the blinds closed, they put me back in touch with reality, reminding me of the incredible possibilities that inhere within my humanity. They tighten that hosepipe connection, restoring my gumption supply so that I can ride on a surge of positive feedback, allowing me to keep those blinds wide open. These meditations are my very personal form of holy communion, where I quiet my yang-self, to see and hear and feel the *real* universe. Through a yielding of the ego, a letting-go of the desire to pander to its insecurities, my yin-self is freed to feel the pull of that unbroken thread which joins us to the ancient beginning, to feel the timeless touch of the *Tao* of wisdom.

GREEN
BOOKS

We publish books on a wide range of Green issues. If you would like us to send you our complete catalogue, please return this card to us. Please also let us know your particular fields of interest – this will help us to shape our future publishing programme.

My particular interests are:

☐ New economics
☐ Wholistic health
☐ The countryside
☐ Organic gardening & farming
☐ Communities & lifestyles
☐ Practical books
☐ Green classics
☐ Green spirituality
☐ Education
☐ The arts
☐ Green philosophy
☐ Green novels & poetry
☐ Green children's books
☐ Global environmental issues

TITLE OF BOOK PURCHASED _____

Do you have any comments on this or any other of our books? If so, please tell us here (or write to us at the address overleaf):

_____
_____
_____
_____

Your name _____

Address _____

_____ Postcode _____

 Printed on recycled material

GREEN BOOKS
Ford House
Hartland
Bideford
Devon
EX39 6EE

6

Being: An Existential Rainbow

TOUCHING the *Tao* of wisdom is not the final step on the trail of spiritual growth. Unfortunately, it is close to being the very first step. The truth is, of course, that our spiritual gate will not stay open on its own. It tends to shut of its own accord. And often we prefer to keep it like that—tightly closed. We choose not to be wise. Wisdom can be dangerous. In the words of Marilyn Ferguson:

> At some point early in our lives, we decide just how conscious we wish to be. We establish a threshold of awareness. We choose how stark a truth we are willing to admit into consciousness, how readily we will examine the contradictions in our lives and beliefs, how deeply we wish to penetrate. Our brains can censor what we see and hear, we can filter reality to suit our level of courage. At every crossroads we make the choice again for greater or lesser awareness.[1]

To take up the challenge of trekking the trail of spiritual growth is to commit ourselves to choosing greater over lesser awareness at every crossroads in our life. It is to set about recovering the self-awareness that we have previously surrendered in defence against the frightening reality of reality. It is to embark on a religious journey of becoming, becoming ever more wholly our self, becoming what we potentially *can* become, as human individuals. The aim of this journey is not to win salvation for our soul, nor is it the attainment of some mystical state of blissful consciousness. The goal is the enhancement of that positive dialectic between love and meaning, to relate ever more closely the inner and outer aspects of our personal reality. This is neither an objective nor a subjective goal, but a particular mode of experience, a certain way of being in the world. Being.

Being is a potential, a capacity, which is written into the very finest patterning of our psychic structure. It is mysteriously encoded within our innate humanity. And, as Abraham Maslow points out in his pioneering work *Toward a Psychology of Being*, as a capacity it is also a need, and therefore an intrinsic value. Just as our physical capacities to see and hear, for example, are transformed into a need to see and a need to hear, so our psychological and spiritual capacities are transformed into needs in the same way. We shall discover that as our highest capacity, Being is transformed into our highest need, and therefore our highest value. This is neither an external nor an internal value. It is value which has to be lived. It can only be actualized dialectically through interacting with our environment. Here is Maslow:

> Man demonstrates *in his own nature* a pressure toward fuller and fuller Being, more and more perfect actualization of his humanness in exactly the same naturalistic, scientific sense that an acorn may be said to be 'pressing toward' being an oak tree, or that a tiger can be observed to 'push toward' being tigerish, or a horse toward being equine. Man is ultimately *not* molded or shaped into humanness, or taught to be human. The role of the environment is ultimately to permit him or help him to actualize *his own* potentialities, not *its* potentialities. The environment does not give him potentialities and capacities; he *has* them in inchoate or embryonic form, just exactly as he has embryonic arms and legs. And creativeness, spontaneity, selfhood, authenticity, caring for others, being able to love, yearning for truth are embryonic potentialities belonging to his species-membership just as much as are his arms and legs and brain and eyes.[2]

The human mind is unique in its plasticity, a quality which has considerable drawbacks in addition to the more obvious advantages. The disadvantage is that we have to learn to do almost everything in life, at considerable cost in terms of time and effort. We remain utterly helpless for the first few years of our life, and, in our society, are not deemed to be a fully independent creature until we have lived for at least sixteen years. However, the advantage is that there is very little we cannot learn to do eventually, if we are prepared to put in the necessary time and effort. Unlike the rest of the animal world, there is always a distinct gap between who we are and who we could be, which is always confused with the gap between who we *think* we are and who we *would like* to be. The degree to which we can achieve a synthesis of these identities is determined by our own individual choices in life. Here is the existential psychologist Rollo May in a passage from *The Courage To Create*:

> In human beings courage is necessary to make *being* and *becoming* possible. An assertion of the self, a commitment, is essential if the self is to have any reality. This is the distinction between human beings and the rest of nature. The acorn becomes an oak by means of automatic growth; no commitment is necessary. The kitten similarly becomes a cat on the basis of instinct. *Nature* and *being* are identical in creatures like them. But a man or woman becomes fully human only by his or her choices and his or her commitment to them.[3]

The fulfilment of our natural potential requires both a fertile environment and a courageous will. The problem that we face today is that our spiritually desolate environment does very little to encourage us to make that choice for greater awareness. In fact it actively discourages us. We are besieged by superficiality. Our culture has developed in such a way as to steer us away from the authentic concerns of life, offering distraction rather than provocation. It has become enslaved to the pseudo-religion of consumerism, serving to shape us according to the needs of the market rather than helping us to shape ourselves according to the real needs of our own potential. Advertising is now our most vital artistic medium, the cornerstone of today's culture, no longer simply displaying and promoting products, but actually generating a need for them, creating artificial dissatisfactions and desires that can only be met by new products. Instead of just defining a product, the images of advertising now function just as much to define the consumer, conferring identity through consumption, selling a product as an integral part of a packaged life-style. In this way, consumerism is sold as a way of life, with the result that reality and illusion have become more and more hopelessly entangled within the consumer consciousness of the modern Western mind. Our genuine needs are jumbled up indistinguishably with those that have been artificially cultivated by the market.

To live authentically in the modern world we have to face up to certain metaphysical questions that were quite unknown and irrelevant to our ancient forebears. Their colourful mythologies endowed them with the belief that they inhabited a magical wonderland suffused with immediate meaning. In contrast, today's forbidding mythology makes us believe that we inhabit a featureless spiritual wasteland. We are told that we stand in bleak and icy isolation before the cold, contingent winds of fate. Our study of the physics of participation (Chapter 2) was undertaken in the hope that it might serve as a sufficiently potent antidote to this invisibly invasive metaphysical poison. I hoped that it would function as an effective vehicle for a protest against the disenchantment from wonder that we presently feel, a signal for a return to some new mystical vision of the universe. What is clear is that the incredible advance in the scope and depth of our analytically derived knowledge has now to be matched by a similar advance in the scope and depth of our intuitively derived understanding. But of course a very real obstacle stands in the way of such an advance, the same obstacle indeed that we have been struggling to overcome through-out this book—the intrinsic limitations of language. To give an authentic answer to our existential question we have to go beyond the dualistic categories that shape our rational thinking. And to do this we cannot rely on words alone. We have to take off on our own and explore experientially.

Although no work of prose can ever hope to answer the ultimate religious question of meaning, this book nevertheless aspires to be of some help by attempting to construct some tentative kind of conceptual framework in which to interpret and map the data of our experience. This is the chief motivation behind our discussion of

Being. Choosing the right kind of language for this job has naturally been problematic. Poetic language tends to be too vague and nebulous. Intellectual language tends to be too formal and rigid. Already I have been guilty of using words like 'spirit' and 'soul' and 'meaning' in far too loose and cavalier a way. There is a difficult compromise to be struck between fluency and accuracy. So far, I have veered toward the poetic use of words, hoping to convey the right kind of intuitive feeling without worrying too much about precision of meaning. Now, I think, it is time to veer slightly the other way and attempt to be a little more precise. However, the fact remains that Being cannot be systematized. This must be constantly borne in mind. To try to frame the non-dualistic concept of Being within the dualistic structure of language is, quite simply, to fail. The unavoidable fact is that, ultimately, Being can only be experienced.

To begin this discussion, I must emphasize again that Being is not some kind of subjective state of consciousness. Being necessarily involves a relationship between the individual self and the world. Being is always Being-in-the-world. Our mind, our body and our environment form the system which is our existence. The description is an awkward one because it is unavoidably dualistic, but Being arises out of a circuit, a complex circular flow of 'information' between mind and world via the interface of the body. In order to get a better feel for what Being is all about, it is vital that we consider the central role played in this dynamic process by the model of the system which is held within our mind, and which, moment by moment, is continually being reconstructed through our perception. It is this internal construction of our immediate reality that is used as the basis for all the mental processing which necessarily sits behind our functioning in the world. The most important element in this model is the image that we hold of our own self. It is this image that we naturally tend to identify with as 'I'. This is our ego, our yang-self, home to our rational, language-based, thinking mind. If, as is usually the case, we have never been led to believe anything different, we naturally come to identify our self-image with the sum total of our personality. We mistake our internal representation of the self for the whole totality of our real self, the Self. Of course, since we can only know this greater self through the 'software' image of it that we create in our mind, it is hardly surprising that we tend to think that this model is all there is to us as a personality. But the ego is *not* the Self. Formed by our thinking perception, the image of personality that we identify with as 'I' is a model in exactly the same way as the colourful image of the world we form through our visual perception.

The structural patterning of the brain is such that our ego can never wholly embrace the full wealth of our humanity, chiefly for the reason that the sub-selves of the Self speak different languages. For example, the ego—our rational self—cannot communicate directly with our intuitive self. Nor can it interface in a straight-forward way with the collective unconscious—our archetypal self. Any understanding of the content of these underground regions of the Self is only ever gained in an indirect way, via the strange medium of dreams, for instance, or through feeling. The intuitive self cannot articulate its solution to a problem in formal language so it adopts a 'getting warmer, getting colder' kind of approach, sending out feelings which enable the rational self to home in on its wisdom. Although the ego cannot directly engage with this intuitive understanding, what is vital is that it acknowledges that it does exist and does have value. Our self-image cannot incorporate a model of the unconscious structures within our Self, but it can incorporate an interface to them, and, just as importantly, a readiness to monitor that interface.

Returning to our system of epistemic feedback around the mind-body-world circuit, I would now like to suggest that the integrity of this system is primarily determined by the authenticity of our mental representations of reality. The better the fit of our internal model of the world—which naturally includes our internal model of our Self—the greater the integrity of our mind-body-world system of existence in other words, the greater the integrity of our Being. The root of the word 'existence' is *ex sistere*, which means, literally, 'to stand out', and that is exactly what our self-awareness allows us to do. It confers existence upon us by making us stand out from the ground of universal Being. In the process, though, it also confers upon us a great deal of anxiety. With our separate mind, our separate experience of being, we find ourselves invaded by a terrible sense of isolation, a feeling that we stand utterly alone, powerless in an infinitely indifferent universe. This acute anxiety, usually denied admittance to the conscious realms of our mind, drives us to try to bridge our great existential divide—to rejoin the ground of universal Being. The only way in which we can effectively achieve this is, very crudely speaking, by 'tuning' our Self to the frequency of the spiritual ground of the universe. Being, in this way, represents a kind of standing out in a way such that there is no apparent discontinuity. We are joined to the ground of universal Being simply because there appears to be no join. Alan Watts offers a beautiful description in *The Meaning of Happiness*:

> The ego and the unconscious, man and nature, oneself and life are seen as the two dancers who move in such close accord that it is impossible to say which moves and which responds, which is the active partner and which the passive.[4]

In this way, Being is our intangible and elusive rainbow bridge between the personal and the universal. This is why it is our highest value, because it is our highest need. Highest here does not mean first or most basic. Being sits at the top of a hierarchy of needs which starts with the most fundamental, like sustenance and shelter and security, goes on to the need to be loved and to be respected, and then reaches a pivotal point where the needs are turned out *to* the world rather than *from* the world, passing through the need to love, and, finally, on to the need to be. These higher needs do not force themselves upon us until the lower needs have been met, and they can recede again when the satisfaction of a lower need is withdrawn or is no longer guaranteed for some reason. The movement up this hierarchy is Maslow's 'pressure toward fuller and fuller Being'. It is this inner force which we should perhaps think of as the human spirit. This is the will to Being. Nietzsche called it the 'will to power'. And power is not an inappropriate term. The experience of Being is indeed an experience of power, an experience of the creative power of the universe flowing, unhindered, unresisted, through mind and body, ego and non-ego in perfect harmony.

However, when the ego bends this power to its own deluded ends, it is far more obviously manifested as a competitive power, the power of the ego pitted *against* the non-ego. The danger for the individual, and now for the world at large, is that this exercise of the will to power has no constraints. It draws power upon itself, always seeking more in an escalating spiral. There is no final satisfaction. This is what is so frightening about the pattern of human life today. We inhabit a world of no restraint; we live under the illusion that there are no constraints. Clearly, the future of our collective human existence on the Earth depends on our ability to develop a pattern of living which, by properly respecting our spiritual potentiality, encourages the authentic exercise of the will to power through Being.

To repeat the words of Susan Griffin quoted in the last chapter, 'we have traded our real existence, our real feelings for a delusion. Instead of fighting for our lives, we bend all our efforts to defend delusion.' This delusion is the misguided belief that the ego is the Self—that the ego is our identity. And it is exactly because we believe that we are defending our very identity that we do indeed bend all our efforts towards the protection of our self-image. By challenging the beliefs which support our ego–identity, change and growth and transcendence pose a threat to the very foundations of our selfhood. We strive desperately to protect this secure and solid picture of our self for the simple reason that it contains our spurious answer to the tragic question posed by our mortality. In effect, we are protecting our self-image from the image of death, and the idea evokes as much passion as the reality. Here is Ken Wilber, from *No Boundary*:

> Man comes to nurse the secret desire that his self should be permanent, static, unchanging, imperturbable, everlasting. But this is just what symbols, concepts, and ideas are like. They are static, unmoving, unchanging, and fixed. The word 'tree', for example, remains the same word even though every real tree changes, grows, transforms and dies. Seeking this static immortality, man therefore begins to center his identity around an *idea* of himself—and this is the mental abstraction

called the 'ego'. Man will not live with his body, for that is corruptible, and thus he lives only as his ego, a picture of himself to himself, and a picture that leaves out any true reference to death.[5]

Again, the root problem is that today's reality is so very much greater in size and complexity and ambiguity than that known to our ancestors, and is considerably compounded by the fact that mainstream Western culture actually offers very little which is of any genuine value in helping us come to terms with what it means to be a human being in this confusing and demanding world. Lacking a coherent mythology, it cannot offer us a coherent sense of identity and purpose. And it cannot offer us help in coming to terms with our mortality. This sits at the very top of a list of socially prohibited topics of discussion. Developing our self-image under the pervasive influence of the Cartesian world-view, we naturally grow up to regard the world as other than Self. We even grow up to regard our own bodies as other than Self. We are conditioned to believe that our mind is separate from our body, sitting on top of it like a pilot. Indeed, for some the mind is seen to be more like a passenger spirit, a disembodied hitch-hiker, taking lifts through one life after another, riding on a succession of different physical life forms. Even within the mind itself this conditioned alienation continues. Unconsciously, our awareness is constrained by powerful social taboos. We learn that we are not supposed to have certain thoughts and feelings—so they are disowned and excised from our self-image. Aspects of our own mind are rejected as other than Self. Consequently, we grow up with a self-image that is a woefully inadequate model of our potentiality as a human being.

At this point, I think it will be useful to recall the classic metaphor that was introduced in Chapter 4, the one that pictured our self-awareness in terms of a window opening out from the fabric of reality. It was suggested then that rather than identifying with the 'physical' body of the window, we should more correctly be identifying with the 'light' that streams through it. This identity may actually be a rather romantic simplification. The considerably more subtle reality is that we should rightly identify ourselves with the light *and* the window. Our being resides somewhere between the personal and the universal, at a paradoxical point which somehow bisects self-consciousness and selflessness.

To elaborate this picture, in the hope of making it a little more accurate—at no little risk of stretching the metaphor beyond its natural breaking point—I would further like to suggest that the window be more realistically likened to a lens, the quality of which is chiefly determined by the clarity of the glass—in other words its relative freedom from cloudiness and aberrations. The quality of our lens in these terms can be understood to be analogous to the integrity of our mind-body-world system of existence. The more closely our internal model of the world is matched to reality, the more closely we live to our full potential, the greater the clarity and the depth of focus of our lens, the more intensely we radiate with the will to Being. The ego-less being of Being can now be seen to be very far from ego-less. Paradoxically, it is only when the ego is strong and well-defined that self-consciousness can be left behind. Only when the ego, to the extent that it encompasses the deeper patterns of our potentiality, is authentically secure in itself, as opposed to in its defences, can we relate purely and completely to the world outside ourselves.

It is the ego's defences that are the major cause of aberration in our lens of awareness. These psychological barriers are fixed opinions which we hold about ourselves and our world, and which serve principally to reinforce the fragile identity of the ego, protecting it from awareness, because any new awareness, always comes packaged with an implicit and threatening call to action. Since understanding carries responsibility, it also brings anxiety and probably guilt. This is the hidden price of new freedoms, and one of the main reasons why our ego seeks to hide from them by repressing knowledge and inventing fictional cover-ups. These counterfeit internal values prevent us from focusing sharply on reality. They blur our vision. Our dynamic reality of process cannot be met with a fixed pattern of ideas like a framework of scaffolding poles. It can only truly be felt with a completely open mind, like a soft, perfectly malleable glove capable of adapting to any situation.

When reality is met with a rigid set of values, the inner shape of a situation cannot be moulded to match its outer shape. It is this dissonance between reality as it is and reality as we have constructed it that interferes with the clarity of the glass in our lens and the integrity of our Being.

To refer now to the metaphor introduced in the last chapter, these aberrations function like restrictions in our gumption supply line, blockages to the flow of our enthusiasm. In *Zen and the Art of Motorcycle Maintenance*, Robert Pirsig calls them gumption traps. He believes that there are two main types: set-backs, which arise from external circumstances, and hang-ups, which arise from internal conditions within ourselves. We all have experience of set-backs, sudden unfortunate incidents which knock us off track and make us lose heart. Although we tend to think we have little control over them, the fact that we experience far more set-backs when we are 'down' would seem to indicate that we are actually far more responsible for them than we might care to admit. When we are low on gumption, we become more vulnerable to the vagaries of external circumstance. The real problem, however, is value-rigidity. Our inability to deal with set-backs is founded on our inability to re-evaluate our internal representation of reality. This is also the root cause of the rather more subtle gumption trap that is an internal hang-up. Again, it is our commitment to rigid values which cuts us off from the awareness that could solve our problems. We get bored, impatient, self-pitying, out of tune with ourselves, out of swing with life, because our values are stuck in a fixed groove.

The first step towards swapping value-rigidity for value-flexibility is the simple recognition of the overblown status of the ego and a consequent willingness to begin a slow process of deflation, a desire to replace the ego's 'hot air' with gumption! Such a step can only be made as the result of both conscious choice and unconscious dynamics. A growing conscious awareness eventually hears an increasingly insistent unconscious voice, and, as if by magic, a trail appears where there was none before. Although we might not at first recognize it for what it is, this is the spiritual trail of becoming. As soon as it is understood that the Self is more than the ego, a new embryonic self-image is conceived, a new tentative centre to the personality. This is a self-image that is willing to open its boundaries to new possibilities, a self-image that actually demands to be allowed to grow, indeed, that *needs* to be allowed to grow.

Unfortunately, the very first section of trail contains many false leads. And it is not at all obvious which is the true trail. We understand the need to turn our attention inwards, to concentrate on our inner life, but a certain degree of self-absorption can easily turn into self-preoccupation. Instead of dismantling the mental scaffolding of our value-rigidity, we merely end up replacing it with an alternative set. These new values tend to be oriented more internally now than externally, but they still serve largely to defend the resilient self-image that is our ego. The true trail of growth extends beyond a second gate of change. Its lock is the hang-up of the ego's fierce autonomy, the key to which is hidden away somewhere in the unconscious. As a result of our growing awareness, both above and below consciousness, which brings us to the right point in space and time and emotion, we begin to cross over a threshold of understanding. Our second gate of change is unlocked from the inside. This is the *birth* of a new self-image, a new, more confident centre to the personality, a self-image with both humility and courage, a self-image with a new epistemology, with an ear for the symbolic voices of its unconscious partners in selfhood. Indeed, this is the birth of an altogether more profound kind of selfhood, an integrated selfhood, secure in its individuality, but none the less aware of its own essential ambiguity.

I was recently privileged to be witness to what appeared to be the very moment at which my wife Shannon crossed over such a threshold of selfhood. We were newly married, travelling in a strange country, sharing in both joyous and traumatic experiences, and consequently getting to know each other in a much deeper way than is possible in the familiar surroundings of home and family. It was a very special and valuable time together, but far from easy. For each of us, for different reasons, it was a time of considerable emotional vulnerability. One particular evening, having been left alone in our shared Moroccan house, we were absorbed in a deep discussion, trying to interpret the bizarre content of a series of disturbing dreams that Shannon had experienced over the previous few nights. Our amateur psychic detective work had thrown up a few likely leads, and we were pursuing a promising new line of investigation, when a meaning suddenly surfaced from the conversation which brought her bolt upright. It was only an intangible idea, but even in my perception, it impacted with a palpable force. Shannon instantly curled up into the foetus position, unconsciously providing a physical metaphor for what proved to be a kind of rebirth. The dialogue between her conscious and unconscious had built to a crescendo. The gate had been reached, the lock located, the key given to her through her dreams. Symbolically, the understanding of the dream's message represented the final turning of the key. To paraphrase Shannon's own words at the time, she killed the old metaphor for her Self, and gave birth to a more authentic one. She felt as though she had suddenly gained the summit of a mountain she had been climbing all her life. Previously, her view had always been impaired by the surrounding hills, but now, for the first time, she could see clear all around. Just a glimpse of that panoramic view was enough to give her a vital outline of a new metaphysical map—the map of her own potentiality. However, as she was soon to realize, having a map of the territory is only a preliminary to the demands of exploring it. A long journey lay ahead—as it lies ahead of us all.

The chief characteristic of this kind of transformation in selfhood is that internal and external values become subordinated to dialectic values. We find ourselves responding to the events of our life on the basis of their explorative value, their creative value, their value in helping us to make actual what is potential in our humanity. Here is Erich Fromm in a wonderful passage from *You Shall Be As Gods*:

> Man is not a subject opposing the world in order to transform *it*; he is *in* the world making his being in the world the occasion for constant self-transformation. Hence the world (man and nature) is not an object standing opposite to him, but the medium in which he discovers his own reality and that of the world ever more deeply.[6]

It is by way of this integrative dialectic that we make authentic progress along the trail of becoming. To take another perspective, this dialectic is a learning process which is equally a process of *unlearning*, unlearning old, previously unquestioned assumptions and habitual responses, old values. Discovering our own reality and that of the world more deeply has a lot to do with uncovering the illusions that we have incorporated into our mental representations of self and world throughout our lifetime, particularly during our earliest, most formative years. To take yet another perspective, it is about curing our hang-ups, bringing who we think we are closer to who we really are.

What causes this disparity is our fear of freedom. At the most fundamental level, our freedom is founded on our imagination and its ability to conceive a variety of future possibilities. We have, potentially, a completely free choice as to which of these futures we wish to try to actualize. However, the very fact that we can visualize a future possibility means that there is a demand placed upon us to change, a demand to move from what we know towards what we don't know. It is both the intrinsic responsibility and the uncertainty of such change which we find frightening—or, more properly, which creates anxiety. In fact, as was first demonstrated so emphatically by the immense creative genius of Kierkegaard, freedom and anxiety are inseparable. Freedom is always freedom from bonds, bonds which provide security. To untie the bonds in our life is to become more completely individual, more alone, more independent, more aware. It is to have to become more self-secure and more responsible—which naturally generates anxiety. However, as Kierkegaard said, 'To venture causes anxiety, but not to venture is to lose oneself.'[7]

Freedom also introduces another peculiarly human kind of suffering, for freedom is never unlimited. Our ability to conceive possibility inevitably means that we also conceive impossibility. Our imagination brings us abruptly into contact with our human limitations. Just as the gap between actuality and potentiality is a source of anxiety, so is the gap between our aspirations and our limitations. The most fundamental of these limitations is our inability to meet in full the contradictory needs of our psyche. Since our different selves each have their own personality, their needs too are naturally different, and frequently oppose one another. Meeting the demands of one self often means neglecting the demands of another. This, simply speaking, is why it is so difficult to find satisfaction in life. When we are alone, one of our selves is wishing for company; when we are in company, another self wishes we were alone. When we are travelling, one of our selves is wishing to be safe and comfortable back at home; when we are at home, another self is wishing for the adventure of travel. In the same vein, in *Toward a Psychology of Being*, Abraham Maslow stresses that psychological growth cannot be considered independently of psychological safety. We are subject simultaneously to *both* a pressure towards growth and a pressure towards safety.

> Every human being has *both* sets of forces within him. One set clings to safety and defensiveness out of fear, tending to regress backward, hanging on to the past, *afraid* to grow away from the primitive communication with the mother's uterus and breast, *afraid* to take chances, *afraid* to jeopardize what he already has, *afraid* of independence, freedom and separateness. The other set of forces impels him forward toward wholeness of Self and uniqueness of Self, toward full functioning of all his capacities, toward confidence in the face of the external world at the same time that he can accept his deepest, real, unconscious Self.[8]

Again, when we are free and independent, one of our selves is wishing for the security of bondage; when we are secure, another self is wishing for freedom. Recognizing that both these needs are authentic, what is important in our healthy development is that there exists a *guarantee* of safety in the form of *authentic* love. We will be encouraged to explore dangerous new psychic territory when we understand that emotional safety is never far away, when we know that we can always retreat. When we have the guarantee of authentic love, the anxiety that comes with growth can be accommodated far more easily.

At no time of life is love more important than during childhood, yet I would suggest, no doubt contentiously, that very few children properly receive the love they need. Of course children normally receive plenty of care, but they often lack the respect, responsibility and understanding that, along with care, form the vital components of *authentic* love. Instead, what they usually get is control disguised as love. Instead of being treated as a unique individual, they are often regarded as an extension of their parents' identity. Love is conditional upon their conformity to their parents' expectations. We can probably all, to some degree, recognize this pattern in our own childhood, especially in regard to emotions. Our parents, thinking they were doing the right thing, taught us what we should and should not feel, telling us that it is bad to have certain feelings, that we mustn't feel this, that we mustn't feel that. From the moment we have our first crude comprehension of language, we learn to stifle and deny our emotions. We learn to disown part of our emotional sensitivity. We also become uncomfortable with our nascent sexuality. We learn to isolate our emotional and sexual selves from our ego.

Such a denial of our true feelings, reinforced not just within the pattern of family life but within the pattern of cultural life too, quickly becomes a habit of mind. We solve the problem posed by the conflicting natures of our different selves by forming a highly selective self-image. We learn to deny fundamental aspects of our humanity. We fail to explore our human potentiality because we have affirmed the need for safety over the need for growth. We have incorporated the need to feel secure into our self-image and erected a defence which keeps the need for growth at bay. We have already explored some of the reasons why this is so. Again, much of the problem is due to the failure of our culture to offer us a way to an authentic sense of identity. Instead of learning to cultivate an identity through our inner potentiality, through the natural expression of our humanity, we are conditioned to manufacture an identity in outer appearances, most commonly through achieving material 'success'—the capitalist touchstone of the 'good life'. Lacking a solid individual sense of our own existence, our identity comes to be rooted in our self-worth, validated externally in terms of what we own and what people think of us. The most important need in life becomes the continual reaffirmation and reinforcement of this derived identity. Our real existential and emotional needs are replaced by the need to be liked and loved by the 'right' people, the need to belong to the 'right' groups, the need to follow the 'right' fashions, the need to believe in the 'right' clichés. We can end up so thoroughly adapted to this social pattern that we can almost totally fail to recognize the demands made by the pattern of our own humanity. And when our self-image, rather than being validated from within by our own human qualities, is validated from without by factors in the environment which are ultimately beyond our control, that self-image is inevitably a very insecure self-image.

It is this failure to be true to our Self, in flight from freedom, that is chiefly responsible for our hang-ups—those internal gumption traps which interrupt the flow of our enthusiasm. If, for reasons of social conditioning or personal insecurity, we are compelled to deny certain feelings or tendencies or capacities, the extraordinary dynamics of the psyche ensure that this knowledge is excised from our self-image. In the language of psychology, the knowledge is repressed. It is not forgotten, though, just locked away in our unconscious, from where it tends to work all manner of devious subterranean sabotage. Our alienated self fights back. It rebels against inanity and spiritual betrayal. Prevented from receiving a conscious hearing, and seeking to voice themselves in some other way, repressed feelings and aspects of our Self can find eventual expression in the body as some kind of tension or illness. Very often, physical symptoms of sickness can be traced back as a psychosomatic response to repression. In fact it is appropriate in some cases to interpret physical symptoms in the body as part of the overall movement towards the health—the wholeness—of the Self. They are our Self's last resort of communication when the ego's defences have made us deaf to all other forms of intuitive communication.

They can also be interpreted as our Self's way of redefining the arena of unresolved conflicts in order that they might become manageable. Illness can be seen as a drastic but none the less effective means of avoiding anxiety. The psychological and physiological dimensions of our self-regulatory organic functioning cannot be considered in isolation. Here is Rollo May, from *The Meaning of Anxiety*:

> Having a disease is one way of resolving a conflict situation. Disease is a method of shrinking one's world so that, with lessened responsibilities and concerns, the person has a better chance of coping successfully. Health, on the contrary, is a freeing of the organism to realize its capacities.[9]

Repressed knowledge is thought to be incorporated into what is called the shadow, an unconscious personality which functions as a repository for all those feelings or tendencies which have proved to be too uncomfortable or frightening to be admitted into our consciousness. Our fierce resistance to these 'alien' aspects of our fundamental nature gives rise to the psychological phenomenon called projection. Because we cannot deny the reality of any part of our humanity, or any experience of our humanity, those aspects that have been repressed as not-Self are projected externally—to get caught on any available hook in our environment. What we most dislike or even hate in other people is often in fact what we most dislike or hate about ourselves, what we have failed to come to terms with about ourselves. There are many classic examples: the person who shows an outraged disgust at homosexuals is invariably someone who has failed to come to terms with their own natural homosexual tendencies; the person who is excessively moralistic is invariably someone who has failed to come to terms with their own natural immoral tendencies.

Obviously, projection represents a gross distortion of our reality—a distortion which has to be incorporated into our internal simulation of reality. Our mental representations of Self and world have to be grossly falsified in order to keep our conscious self blind to what is going on behind the scenes. By repressing facets of the totality of our person, we also repress their original meaning for us, and so lose touch with our deepest motivations, which are usually replaced by personal or cultural clichés. Such misinterpretations— like, say, the substitution of the need for physical closeness and intimacy with the more superficial need for sex, particularly by men—have the effect of making us incomprehensible to ourselves.

For example, we can get stirred into a self-devouring stew of frustration, bitterly resenting the demands of our environment, blaming other people, usually those closest to us, for our discontent, when, in reality, that discontent is totally self-conceived. The source of our resentment is invariably to be found within our own shadow. In this way, we invent our own obstacles. We restrain our own capacities from within. We fight against ourselves. As Ken Wilber says, 'It's a boomerang effect, and you end up clobbering yourself with your own energy.'[10]

Repression and projection are just two examples of the many ways in which our self-image defends itself. It is beyond the scope and need of this book to discuss the process further, for it is the principle that is most important here. First and foremost, defences are constructed as a protection against anxiety—the anxiety that accompanies freedom. We shrink our vision of our Self so that we don't have to face up to the more disturbing aspects of who we are, or the frightening challenge of having to develop into who we potentially could be, and having to bear all the responsibility that that would inevitably bring along. However, by shrinking our self-image in this way, we entrap ourselves in a vicious circle. Distorting our internal model of the world in order to restrict our possibilities also restricts our choice among our remaining possibilities. Because our awareness is filtered in order to keep our illusions veiled, to preserve the viability of our distorted images of Self and world, so our choice of possible futures becomes restricted to those that will not threaten our system of beliefs, those that will not undermine our personal mythology of delusion. Consequently, our self-image shrinks even more, further diminishing our freedom. We may think that we are always acting freely, of our own volition, but we are not; uninformed by our true motivations, we are more and more to be found behaving rigidly in response to our hang-ups about ourselves, living out habitual patterns dictated to by the fixed programming which has been coded into our unconscious mind, unbeknown to our ego. We are never fully in control of our life because there always exists the possibility of being held hostage to these hidden forces, the possibility of being hijacked by our delusions.

We make progress along the trail of becoming by facing up to our freedom, by attempting to uncover those aspects of our Self which we have previously deposited in our unconscious. In order to cure our hang-ups; in order to see through the illusions that we have patched into our reality simulations; in order to unlearn our false assumptions and redundant values, we must, in Ken Wilber's words, re-own our projections:

> To re-own a projection is to dissolve a boundary. When you realize that a projection which appeared to exist 'out there' is really your own reflection, is actually part of yourself, then you have torn down that particular boundary between self and not-self. Hence the field of your awareness becomes that much more expansive, open, free, and undefended. To truly befriend and ultimately become one with a former 'enemy' is the same as tearing down the battle line and expanding the territory through which you may freely move. These projected facets will then no longer threaten you because they *are* you.[11]

There are three vital prerequisites for this kind of self-exploration: humility, courage, and a discipline of mind which forces us to look beyond simple and convenient explanations. For as Joan Borysenko states so precisely in *Minding the Body, Mending the Mind*;

> We rarely see things for what they are. Instead, we see the reflection of our own conditioning.[12]

The clouds of our self-delusion serve largely to protect our ego, which will always prefer to make excuses rather than face up to its shortcomings. It is all too easy to avoid asking ourselves the most difficult questions, and all too easy again to avoid answering them honestly. The honest answers are not usually the answers we want to hear. Ultimately, one of our most useful navigational aids on the trail of becoming is our own dishonesty to our self. Self-deceit can be found at work in countless ways within the psyche. It works, for example, to disguise deeper anxieties so that we are unable to recognize them for what they really are. Often, they are only revealed to us in the form of our inhibitions. Inhibitions are like chinks in the armour of our ego. They allow us to sneak a glimpse at our real anxieties in life, revealing the more vulnerable self that lives behind the social mask we all wear. In this way, through an understanding of the mechanisms by which the ego defends itself, we can monitor our perception and watch out for tell-tale signs like that of projection. Gradually, we can awaken ourselves to some of the skulduggery that is going on behind our conscious awareness. Gradually, we can learn to reown our natural feelings and potentialities, and go on to reown our body, and even reown our world, a world that, very early on in life, we were taught to disown as other than Self. The trail is steep and rugged, but we persevere because the journey we are making is self-validating. As we form a more authentic self-image, so we feel more alive, more genuinely enthusiastic, more intoxicated with the spontaneous, creative power of Being. Our effort is validated by the simple enhancement of our experience of being alive.

The overriding message that I have been wanting to get across here is that we *do* have control over our own programming and that, with effort, it is possible to reach in and debug our own software, online so to speak. It is we who create the reality that we experience, and we hold within us the capacity to enhance the quality of that experience. We are not so much a fixed, static identity, but a dynamic, fluid process, a process with an inner shape and an outer shape, both undergoing incessant change and in constant need of matching with one another. We can never be still. We always need to be responding, developing, growing. And our restlessness will not tire us because as our awareness is intensified, so we find that our creative energy is intensified in tandem. Using this energy with humility and courage, we can grow into the unknown in spite of anxiety. We can grow to become who we truly are.

However, spiritual growth cannot realistically be achieved by shutting ourselves away from the world to enter into full-time self-analysis. Despite the clichéd images, the spiritual trail cannot be trekked sitting alone in some remote and isolated hermitage. We only understand our inner world through the outer world of experience, which in turn allows us to understand the outer world through the inner world of ideas. It is this dance of experience and thought, the dialectic interplay of intuition and reason, that provides the metaphors with which the world is made intelligible. Indeed, the spiritual quest, just like the scientific quest, is metaphorical. The deeper our reservoir of metaphor, the deeper our wisdom, the deeper our understanding of ourselves. We can only acquire these metaphors through engaging with the twin worlds in which we live—the world of nature and the world of human culture. Without human culture, nature is nothing more than a source of food and shelter; without nature, human culture is nothing more than a source of distraction. As mirrors in which to look for a reflection of our essential humanity, neither, on its own, can offer a complete image. We have to flip between the two. It is only by engaging in a dialectic between the interior universe of ideas and the exterior universe of natural experience that we can put ourselves in touch with the deeper aspects of our potentiality.

It is through this dialectic that we are able to increase the depth to which we can focus our lens of awareness. By developing an increasingly authentic self-image, we find ourselves seeing through the false appearances, the illusions, the masks, to the naked reality of things, no longer seeing only what we want to see—and what society wants us to see—but seeing in a universal sense, without the imposition of the bias of our conditioning. This greater, more authentic awareness brings us into contact with the inescapable uncertainty and impermanence of our existence. Understanding that our ultimate questions of life can have no definitive answers, understanding that eternity is only to be found in the ever-transforming moment of the 'now', we learn to accept that we must live—with anxiety—into the unknown and the unknowable. We come to appreciate that uncertainty and impermanence are the very essence of the process of creative renewal that is life. Uncertainty and impermanence stand for the open-endedness that makes life possible—and meaningful.

All living forms, rather like all pieces of music, have a natural rhythm and timeframe. Each living act, like each musical note, has its place within the whole—a whole which is completed by the closure of death. The compositional development of each individual human life is quite unique, with its own special melodies and syncopations. This development may demand a slow final movement, or possibly a crashing crescendo. It all depends on the person and the particular pattern of their life. What is important is that the natural tempo corresponding to this pattern is not violated. Like a piece of music, the pace of life cannot be forced or slowed artificially. To do so would be to injure the harmony of the whole. Those who try to resist the natural tempo of their musical development through life invariably succeed only in transforming their melody into a dirge.

Probably, then, the most important aspect of our humanity that we can bring into full conscious awareness is the fact of our moribundity. Our impending death is an absolute truth that faces us all. It is vital for our full psychic health that it is accepted without resistance—which does not necessarily mean living at peace with it. Rather, it means not ignoring it, not supressing our knowledge of it, facing up to it honestly, without pretence, without the resistance of false hopes and beliefs. Such an active acceptance of our temporality gives our existence immediacy, a more deeply urgent reality. Indeed, it can come to stand as the central creative dynamic in our life.

What if, as George Gurdjieff once conjectured, we possessed an organ which foretold the precise date on which we were going to die? Would this make a difference in our lives? The answer, very probably, is yes. This constant reminder of the exact finitude of our existence would likely provoke us into leading a far more purposeful life. Drawing closer and closer to our predetermined doomsday, we would want to live every moment to the full and, I would like to think, not just selfishly. Material possessions would lose all significance for us. We would want to do something good in the world so that our only too brief little life might be of some greater significance. Why is it, then, just because we don't know the precise date of our death, that we generally live so purposelessly and apathetically? The answer, quite simply, is that because it has never been discussed, because we have learnt not to think about it, death has been deprived of its reality for us. Our lives have no immediacy. I am not suggesting that we go around morbidly contemplating our death all the time, only that we *anticipate* it and incorporate this very concrete, unavoidable eventuality into our self-image. In this way, our imagination can be used partly to simulate Gurdjieff's premonitory organ and so impart a vital commitment into our lives—a commitment to the realization of our human potentiality. Only in this way, having lived the very best life we were given to live, will we be able to give ourselves over to life's need for renewal without any bitterness or regret, in obedience to the natural musical conclusion of our personal symphony of Being.

The acceptance of the impermanence of life is very closely bound up with the acceptance of the uncertainty of life. Both represent the discovery of new freedoms, vital to the vigour and perspicacity of our awareness, and therefore vital to the integrity of our Being. Such an acceptance represents the clearing of a kind of vague fogginess in our lens of awareness. Living at peace with uncertainty means that we can explore the ultimate questions of life with complete openness, unpressured by any need to find answers. This, in turn, means that our curiosity will never be extinguished because the questions will always remain alive for us. Coming to terms with the intrinsic uncertainty of our existence means understanding that life cannot be answered with thought alone, through faith or metaphysical belief. Although we still develop a personal mythology, we know that this cannot, in itself, be our answer to life. We understand that we have to *live* our answer to life, through humility, courage, commitment and ultimately action. We understand that the only truth in life is the truth we live. Ultimately, our faith has to be placed in our own being in this world. We each have to address ourselves to the most basic fact of 'I am', and the question of what we are going to do about it.

The answer to this question which has emerged from every spiritual quarter of the world is love. Learning to let go of fear, learning not to be afraid of who we are, learning to live to our full potential: all are lessons in the art of love. In Chapter 4, we proposed a definition of love largely derived from that given by Erich Fromm in *The Art of Loving*. Love is the conjunction of care and responsibility guided by respect and understanding. Before we go on, I think it will be worthwhile consolidating this understanding a little. First, it has to be stressed that love has no real meaning outside action. Love is an activity, a creative activity. Erich Fromm rightly says that there is no such thing as 'love'. Love is not a thing. We cannot *have* love. There exists only the *act* of loving. Love is indeed an art, a dynamic art that can never truly be mastered or perfected. Like any artistic apprenticeship, an apprenticeship in the art of love is an apprenticeship for life.

We have to be very cynical about the quality of much that passes for love in our culture. What is generally called romantic love is not love in the sense of the word that we have developed here. Without respect and understanding, romantic love is extremely fragile. The outer shell remains intact, but inside, care and responsibility can easily be replaced by power and possession. Disorientating, exhilarating, irresistible, romantic love is nothing short of intoxicating. It is beautiful, and is to be enjoyed to the full, but it is often a chimera and invariably ephemeral. For most people, the ideal ends up being traded for an illusion. Unconditional love is compromised by what Balzac called 'a monster that devours everything'—familiarity.

Love is an art requiring constant practice and concentration. It is not something that we can just sit back and enjoy like a hot bath. Love can only be authentic in as much as it flows out of Being, out of true independence and individuality. Love and Being are always to be found together. Love flows irresistibly out of our authentic Being-in-the-world. It is a function of our self-integrity. It is the active expression of a self-image which is in tune with its own hermeneutics; which, in other words, is informed by an authentic self-interpretation. When we understand our own deeper motivations we are free to recognize ourselves for who we genuinely are, and are therefore able to recognize others for who they genuinely are. The less our communicative interchange is dictated by the hidden agendas of our shadow, the greater the possibility of true intimacy in our relationships—the kind of intimacy in which we are free to touch the inner person as well as the outer, a spiritual intimacy in which we are able to bring each other more fully alive with Being, our capacity for authentic love no longer hung up by our hang-ups.

Having established the very close relationship between Being and love, I think we can now restate more precisely a proposition that was introduced in Chapter 4. *Being and meaning are the objective and subjective poles of the same basic experience.* Meaning inheres in the simple experience of our own aliveness. It is, to borrow Ken Wilber's phrase, to be found in the inner radiant currents of our own being, and in the release of those currents to the world. This, again, is why Being stands as our highest value—because it turns out to be our holy grail, the answer to the ultimate human quest for meaning, our rainbow bridge between the personal and the universal.

In this chapter and Chapter 5, we have been discussing the principal factors, both positive and negative, which determine the integrity of our Being. Making use of our classic metaphor again, Chapter 5 discussed the dynamic, functional aspects, suggesting that our lens of awareness possesses a kind of shutter which has continually to be held open through an effort of conscious will, a will that is fuelled by meaning. In this chapter, we have been more concerned with the structural aspects, the factors which determine the clarity of the lens itself, and its depth of focus. Clearly, the intimate relationship between Being and meaning suggests that the functional and structural aspects are inextricably tied together.

Being is a variable. Trekking the spiritual trail of becoming does not mean that the integrity of our Being increases in a nice linear progression. It means, rather, that the cushioning is taken out of our bumpy ride through life. Where once the ruts were smoothed out by virtue of an in-built suspension system, we are now able to *feel* the ride. And we are rewarded with moments of ecstasy, moments of creative power, where our inner currents of being resonate deeply with the outer currents of the Being of the universe. However, without the buffer provided by that soft suspension system, we also start to feel the full impact of life's potholes. Facing up to the bare reality of our limitations, facing up to our utterly inescapable dependence upon the universe, facing up to the unavoidable contradictions that war in our soul, facing up to the responsibility and anxiety of growing to be who we are; all can thwart even the most courageous of hearts, causing that shutter on our lens to close over. Being and becoming share in a complex dialectic relationship, and are always set against the context of our immediate environment. Our effective capacity for Being is always liable to be compromised by circumstance.

The multi-faceted, open-ended nature of our human personality condemns us to a life of unrest. We are destined always to be striving to balance our conflicting priorities. We have been both blessed with a sense of adventure and cursed with an addiction to boredom. We are creatures of the imagination who are not entirely at home in the world of the imagination, creatures of the Earth who are not entirely at home in the world of the Earth. We always have to act on the basis of incomplete knowledge, yet the responsibility we bear is never less than total. The fundamental problem is that our inner resources are too feeble to meet our godly aspirations. Our consciousness is stricken with design flaws in its underground service systems. To take to the trail of becoming is to enhance the quality and authenticity of our every experience of life—which means that we will grow more acutely aware of our innate design faults, and the fact of our imperfectibility. As we discover new freedoms, so we become increasingly aware of the limitations which contain those freedoms. However, we also become aware that those same limitations are essential to growth. It is our limitations which challenge us to grow. It is our limitations which provoke us to venture creatively into the unknown. As Rollo May says in *The Courage to Create*, creativity *requires* limits: 'The creative act arises out of the struggle of human beings with and against that which limits them.'[13]

Our understanding that we cannot have Being without limitations, nor creativity without conflict, allows us perhaps to live more peacefully with all the skirmishes that break out in our soul. The battle for supremacy among our inner oppositions is written into the unnegotiable terms of our contract of existence. We cannot be true to our Self until we unconditionally accept the terms of that contract. We must accept the anxiety of authenticity and move forward in spite of it, to give form to the universe of creative possibility that inheres within the soul of every one of us, to integrate who we *seem* to be and who we *can* be into the same authentic concept of identity. To be true to our Self is, once more, to choose greater over lesser awareness, and to accept the greater responsibility with which that greater awareness always comes packaged.

As our rainbow bridge between the personal and the universal, Being carries with it an awareness of our global responsibility. Developing an authentic internal simulation of reality means enlarging our identity to include the world. It means identifying not just with an integrated mind and body, but with the whole mind-body-world system. It means expanding our sense of Self to acknowledge the totality of our embeddedness in the whole universe of existence. In our Western tradition, this spiritual embrace of the world might well be interpreted as a love of God; not the thought experience of love which amounts to no more than a belief in God, but the feeling experience of love, the agapeistic love in which we intuit our oneness with God, the God of universal Being, the God that is Mind—the mind-at-large in which we are each individual sub-minds. A growing respect and understanding of the Earth as the fount of our being will guide an increasingly urgent desire to respond to the Earth with care. Only when we feel just this kind of elemental connection to the Earth, when the world is felt, emotionally, to be our world, will we be urged to make the religious commitment that is essential to the preservation of the integrity and diversity of our planet's ecology. This is our hope for the future. Through our awareness of its reality as both the subject and object of our authentic love, we have to rediscover the sacredness of the Earth.

7

The Mythology of Progress

OUR PERCEPTION of progress has been shaped by modern culture to mean just one thing: material progress. It is accepted, unanimously in the case of those people and institutions which are in positions of power, that progress stands for economic growth and technological advancement. Progress has become synonymous with industrialization. But this is to view progress in a very restricted, single-dimensional way. Progress is truly multi-dimensional, with many interdependent contexts. What is progress in one context may not be progress in another. What is progress for one community of people may not be progress for another. Nowhere can this be more clearly seen than among the Earth's aboriginal peoples. For them, Western progress has been catastrophic. Originally condemned by our civilization as primitive savages, their populations and cultures have been almost obliterated by our fatal designs and diseases. Only now, when it is far too late, are we beginning to respect the wisdom and dignity of these peoples. Only now do we acknowledge the possibility that their spiritual lives, as revealed through their stories and myths, are tapped far more deeply into the divine than our own.

Alongside today's cult of technology has grown a cult of explanation. I have previously called it the cult of trivialization. Just as technology has eroded the strength of our relationship with our land, so explanation has eroded the strength of our relationship with our great mythologies. In our knowledge-based society, myth has indeed been debased through trivialization. Saturated by a diet of the profane, our collective imagination has become divorced from its environmental and mythic substrate, disengaged from its two greatest experiential sources of wisdom. Estranged from the landscape of nature, we feel no bond with the Earth. Estranged from the landscape of myth, we feel no bond with our own human dignity and potentiality. The cult of explanation has ridden roughshod over our spiritual aspirations. Nowhere is this more beautifully expressed than in *The Heart of the Hunter*, Laurens van der Post's compassionate tribute to the spirit and imagination of the Bushman people of Southern Africa's Kalahari Desert:

> We know so much intellectually, indeed, that we are in danger of becoming the prisoners of our knowledge. We suffer from a hubris of the mind. We have abolished superstition of the heart only to install a superstition of the intellect in its place. We behave as if there were some magic in mere thought, and we use thinking for purposes for which it was never designed. As a result we are no longer sufficiently aware of the importance of what we cannot know intellectually, what we must know in other ways, of the living experience before and beyond our transitory knowledge. The passion of the spirit, which would inspire man to live his finest hour dangerously on the exposed frontier of his knowledge, seemed to me to have declined into a vague and arid restlessness hiding behind an arrogant intellectualism, like a child of arrested development behind the skirts of its mother.[1]

Reading *The Heart of the Hunter* is a very humbling experience. Despite the undeniable hardship of their nomadic lifestyle, I cannot help but envy the Bushmen their 'natural aristocracy of spirit', as Laurens van der Post calls it. Their spirituality was so rich, so intense, indeed, so natural. And this is not just a nostalgic, romantic longing. The Bushmen may not have had rational explanations, but they did have an intuitive understanding of life which today we have almost completely lost. Here is van der Post again, in another poignant and poetic passage from *The Heart of the Hunter* :

> Intellectually, modern man knows almost all there is to know about the pattern of creation in himself, the forms it takes, the surface designs it describes. He has measured the pitch of its rhythms and carefully recorded all the mechanics. From the outside he sees the desirable first object of life more

clearly perhaps than man has ever seen it before. But less and less does he experience the process within. Less and less is he capable of committing himself body and soul to the creative experiment that is continually seeking to fire him and to charge his little life with great objective meaning. Cut off by accumulated knowledge from the heart of his own living experience, he moves along a comfortable rubble of material possession, alone and unbelonging, sick, poor, starved of meaning. How different the naked little Bushman, who could carry all he possessed in one hand. Whatever his life lacked, I never felt it was meaning. Meaning for him died only when we bent him to our bright twentieth-century will. Otherwise, he was rich where we were poor; he walked clear-cut through my mind, clothed in his own vivid experience of the dream life within him.[2]

Laurens van der Post's respect for the Bushman is almost overwhelming. In the introduction to *The Heart of the Hunter* he quotes from a poem of D.H.Lawrence. These three lines say it all:

In the dust, where we have buried
The silenced races and their abominations,
We have buried so much of the delicate magic of life.[3]

When we have indeed buried so much of the delicate magic of life we cannot possibly hold up our mythology of progress as the definitive mythology. We can lay no such sole claim on progress. Indeed, progress is empty of meaning without the focus of some well-defined goal toward which social change is to be directed. A mythology of progress effectively amounts to a mythology of value. The direction in which a society wishes to take its people is intimately tied to the values it wishes to affirm for its people. The authenticity of progress, therefore, is determined by the authenticity of those values.

In order to explore this idea in detail, I want to discuss a pair of concepts from a very remarkable book, James Carse's *Finite and Infinite Games*. The concepts I refer to are those of the title. Carse introduces them by stating that there are just two fundamental kinds of game:

A finite game is played for the purpose of winning, an infinite game for the purpose of continuing the play.[4]

Similarly, there are basically just two ways in which we can *play* a game: we can either play to win or we can play to continue the play, for the sheer pleasure of the game. These two modes of play are, to a considerable extent, mutually exclusive. And they parallel very closely our two principal modes of existence. Playing to win is synonymous with living to *have*. Playing solely for the pleasure of playing is synonymous with living to *be*. The trouble is that in today's acquisitive, highly competitive society, when everybody else seems to be playing to win, it is not easy to play for the simple joy of playing the game. The joy of playing the game of life has become subordinated to the acquisition of the property which marks each little finite victory on the way up the ladder of success. It is because modern society is so comprehensively geared to the recognition and reward of success in these power games that it is so difficult to play purely for the joy of playing. As a matter of social convention, we are frequently virtually compelled to play to win.

The essence of all these social games is role-playing. It is implicit in their unwritten laws that we must hide our real self behind a mask—our persona, a term which owes its origin to Jung. We are required by social necessity to present the 'right' image—to a prospective boss in the job game, to a prospective husband or wife in the marriage game, to a prospective customer in the marketing game. The more skilful we become at presenting the appropriate image for the appropriate occasion, the more socially acceptable and successful we become. The development of such a persona—as a kind of social interface—is essential to a healthy life. It is a vital part of our total personality. But it should always remain *just* a part—for as we become more and more skilful in our chosen roles, there is a tendency for us to begin to deceive ourselves. We become less and less able to distinguish between the masquerade and the reality. The persona can take over our personality. Our mask can become so convincing that even we end up believing in the image that it represents. Instead of living our own life, we are subtly coerced into living a life-style—or would it be better to say that we are actually lived *by* our life-style?

However, by staying alert to the fact that our social games *are* charades, we open our eyes to the possibility of playing the game of life in a very different way—as a creative game, as part of a universal game, an infinite game. By penetrating to the reality behind the masks, we find that we can draw upon a far greater range and depth of character. Taking part in the charades in the full knowledge that they are only charades, we cannot help but play for the joy of playing because success is no longer dictated by the result of the game, only by its quality. We become alive to the essential paradox of victory—that the winning of a game represents the loss of that game. The winning of a game signals the end of the game. This is not the aim of universal play. Playing for the joy of playing means playing for the sake of the game, playing to bring our 'opponents' into the game, playing to continue the game, to enliven the game, to develop the game. The more we play for the joy of playing, the more possibility we see for future play. The game of life becomes a game of unlimited possibility. By discarding image and ideology, we open ourselves fully to the pattern of our innate potentiality. Although we respect the fact that this pattern imposes limitations, we never submit to any rigid evaluation of exactly what they might be. We incorporate them into our play. They become part of the game. The greatest joy of the game of life is to be playing at the frontiers of our potential, playing not *within* our limitations but *with* our limitations.

Finite and Infinite Games caught my imagination because the concept of infinite play seemed to capture the essence of spirituality perfectly, and to offer a way of understanding how it is to be joined into the social arena. Quite simply, to be spiritually aware is to be alive to the universal, infinite nature of the game of life. Written in a very idiosyncratic style, exploring the nature of finite and infinite play with stunning originality and erudition, Carse's work abounds with aphoristic delights. They leap out at you from almost every page. Many of them make a pointed but subtle distinction between the characteristics of the finite and the infinite player. This is one of my favourites:

> Let us say that where the finite player plays *to be powerful* the infinite player plays *with strength*.[5]

The kind of strength that he is referring to here is, I believe, synonymous with the *Te* of the *Tao Te Ching*. The spiritual languages of Lao Tzu and James Carse overlay quite perfectly. Power requires the context of a finite game. It is dependent on a fixed set of rules and the participation of a certain number of players playing by those rules. If the players decide to change the rules, or decide to opt out of the game, the power of the game-master evaporates. Finite players, then, can always be deprived of their power, and can therefore never be secure in it. Infinite players, however, can be secure in their strength because it is not context-dependent, except in respect of the game of life itself. Infinite players can never be deprived of their strength.

> Power is finite in amount. Strength cannot be measured, because it is an opening and not a closing act. Power refers to the freedom persons have within limits, strength to the freedom persons have with limits.
> Power will always be restricted to a relatively small number of selected persons. Anyone can be strong.[6]

I contend that our present-day determination of progress is unauthentic because it affirms the finite value of power over the infinite value of strength. Grounded solidly in the old Cartesian perception of reality, the Western mythology of progress is written in the language of the finite game. Its simplistic goal is to increase society's prize pool to a level where every player can be a 'winner' in material terms. Its values are dishonest because such a goal is demonstrably unsustainable in ecological terms, and because any such competitive framework will inevitably polarize society between the deeply divisive forces of greed and need. Simple logic dictates that every player cannot be a winner in such a game. The concept of winner cannot be divorced from the concept of loser. There is simply no getting away from the fact that one of the many waste products of a society founded on the principles of finite play is waste people, the down and out losers of its finite games.

The central tenet around which today's mythology of progress has been constructed is economic growth, a principle which has come to assume the mantle of a received dogma. Ideology demands that each nation must strive to increase its material production and consumption—and hence its pollution and waste—in order to create the wealth needed for its progress. Ideology demands that each national market must grow in size, which demands, in turn, that marketing must grow in its aggressiveness. The power of the media must be used to brainwash people into buying products that they previously had no idea they needed so that more and more people can be kept occupied producing yet more new products which no one really needs.

This 'system' exerts an almost totalitarian rule over the patterns of modern life. Typically, the Western imagination, suffocated by an infatuation with having, has degenerated into voyeurism, losing its desire to venture and wonder and embrace its landscape, to experience and quest after Being. It invariably never rises above being a tourist in its own backyard, detached and indifferent, a casual onlooker. Our spiritual eyes have been forced shut, blinded by the dazzling technological glare of all the countless material temptations that we now find placed before us. The joy of infinite play has been scorned in favour of the material rewards of finite play.

Technology released us from bondage to the land, only to deliver us into the bondage of the 'system'. It was originally thought that technology would free us to develop our full human potential, to be creative, to lead a meaningful, spiritually fulfilled life, but although many people have indeed been emancipated, the vast majority have not. The great bulk of humanity has no more time now for genuinely productive, creative activity than they have ever had. In *Small is Beautiful*, E.F.Schumacher admits that he has been tempted to formulate the following proposition as the first law of economics:

The amount of real leisure a society enjoys tends to be in inverse proportion to the amount of labour-saving machinery it employs.[7]

It is a thought-provoking point. In the few non-industrialized areas of the world which are not ravaged by drought and oppressive governments, it is difficult to deny that people do have far more time to enjoy themselves. As Schumacher puts it, 'The burden of living rests much more lightly on their shoulders than on ours.'[8]

A lot of the trouble has to do with the abysmally boring or highly stressful jobs that many people are obliged to do in order to earn money. The mind is switched away from its creative centre by the crushing routine of work—and is never switched back on again. The simplest and most wonderful enjoyment of life, the purest celebration of our humanity, is to be creatively and productively engaged with mind and body—as in drawing, painting, playing music, dancing—yet today we often cannot find the time, or the will, or simply the opportunity to develop our natural creative talents. Here is Schumacher again:

The type of work which modern technology is most successful in reducing or even eliminating is skilful, productive work of human hands, in touch with real materials of one kind or another. In an advanced industrial society, such work has become exceedingly rare, and to make a decent living by doing such work has become virtually impossible.[9]

And again:

Today, a person has to be wealthy to be able to enjoy this simple thing, this very great luxury: he has to be able to afford space and good tools; he has to be lucky enough to find a good teacher and plenty of free time to learn and practise. He really has to be rich enough not to need a job; for the number of jobs that would be satisfactory in these respects is very small indeed.[10]

More than any other factor, it is the vast scale on which modern societies operate that has done most to cut us off from the natural values that were once lived out by our aboriginal peoples. The intricate complexities of their interdependent economies, the convoluted logic of their internal politics, the labyrinthine impenetrabilities of their bureaucratic machinery, the fact that the policy-makers and decision-takers are desk-bound, as good as a million miles from the people their policies and decisions most directly effect: all combine to confer a sense of anonymity upon the individual. Our bureaucratic structures are now so huge and the control of their organization is now so remote that most of us feel no genuine sense of participation in our society. And, for increasing numbers of people, this lack of authentic involvement applies to their jobs too. Work is losing its intrinsic value. Fewer and fewer of us are able to use our hands creatively, and those that do earn little respect or prestige. That applies also to people in the caring professions. Nurses and teachers, for example, undeniably the most important skilled workers in any civilized society, are among the more poorly paid. The really prestigious jobs are those that are furthest from the sharp end, most removed from the scene of actual production and one-to-one care. The very highest paid people are those who work in the most abstract, artificial and humanly meaningless field of all—the financial markets. These are the people who make money *out of* money, although, of course, nothing is actually made at all. All they are really doing is reorganizing the electronic information that is stored on the computers of the world's financial institutions. Their supremacy in the salary league is perhaps a rather fitting symbol for our times.

The sad fact is that our perception of the value of work has come to be shaped more and more by the amount of hard cash that it puts in our pocket. The size of our salary quantifies our success. It has become the most important value in our life. Indeed, it is often so important that all deeper values and principles are sacrificed to this one external value of self-worth. For the sake of our job security and promotion prospects, we have little choice but to surrender our own personal values to those of the 'system'. This is where we have been led by the competitive individualism of our society. We are motivated far more by the value of our pay cheque than by the values inherent in our own potentiality. The natural priority has been turned upside down. Work has become life-consuming rather than life-affirming.

In any short discussion of issues of this scale and complexity it is impossible to avoid falling prey to the twin evils of over-generalization and over-simplification. By concentrating on the failings of our mythology of progress, the picture presented here is inevitably one-sided. We should be very proud of much that our society has achieved. We should certainly be very proud of today's high material standard of living, and be very grateful for it, and enjoy it, but we *must* recognize that there is a price. What is really frightening is our apathy, the fact that so much is now taken for granted. We've been spoilt. We want all the benefits of our kind of progress, but we don't want to bear any of the responsibility for its hidden costs. While our material expectations have risen ridiculously high, our spiritual aspirations have sunk ridiculously low. I believe we have already passed a critical turning point where, although our standard of living—in objective, economic terms—is still rising, the quality of our lives—in thoroughly subjective terms such as peace of mind and well-being and community spirit—is actually declining. It is not something that can be proved by statistics. It is a gut feeling, an emotional feeling which is being shared by more and more people. By enshrining the profit motive as an inviolate governing principle, Western progress is not only bringing about the destruction of the fragile ecology of our planetary home; it is also creating an ever more despiritualized, demythologized, dehumanized society. Progress, judged by the values of many people, has now become regress.

However, even if it could be argued that the quality of life in our society was improving, we would still have to face the appalling reality of the global context in which our 'progress' has taken place. If an extraterrestrial visitor were to visit our planet it would be likely to be very confused and saddened by the state of human affairs on Earth. It would find that the havoc being wreaked on the integrity of the planet's life-systems is not directed at providing an acceptable quality of life for its five billion human inhabitants, but is geared to providing a relatively small, privileged portion of the population with a plethora of material things that they don't actually seem to need, nor make use of in any creative way. Our visitor would find that while this affluent minority is being plagued by the modern 'diseases' of stress and boredom, hundreds upon hundreds of millions of people are suffering from malnutrition and are forced to endure conditions of utter, abject squalor—conditions, strangely, from which they still seem able to conjure up the experience of more genuine joy and dignity than many of the rich people. And it would find that the nations in which these indescribably poor people live are actually in debt to the nations in which the rich people live; so much in debt, in fact, that their entire economies have had to be distorted in order to service just the interest on the loans so enthusiastically offered to them by the rich nations—for development projects, incidentally, which were often disastrously inappropriate to their needs.

Our visitor would find that the Earth's magnificent and desperately valuable tropical rainforests are being destroyed, not for essential domestic food production, but to earn foreign exchange to pay back the rich nations. This is what the poor nations must do if they are to ensure that they can obtain the further loans which they need in order to develop their industrial economies, the only way—according to the rich nations—by which they can climb out of the economic hole which those same rich nations originally helped them to dig. Our extraterrestrial visitor would be horrified to find that this situation was perfectly acceptable to all but a mere handful of people in the rich nations. It would find that the great majority are willing to elect leaders who are utterly indifferent to the massive injustices of their planet—and, worse, who advocate policies of unmitigated lunacy, policies which threaten the very future of the human species. Our visitor would be nonplussed to find that the majority of people in the rich nations choose to elect leaders who support the holding of utterly absurd numbers of nuclear weapons—more than enough to destroy a Hiroshima every second for two whole weeks, continuously, day and night. Just take a minute—or sixty annihilated cities—to think about what that means. If our visitor did, it would decide to return home.

What has happened is that as our original small and isolated communities have combined and grown into larger and larger societies, so our control over our own destiny has gradually been surrendered. We have been taken over by our own ideological systems, the greatest of which is the capitalist system. The tide having now turned against communist ideology—which never amounted to anything more than state-controlled capitalism anyway—the market-controlled capitalism of the West looks set to draw even more power to itself in the next decade. In this system, humanity has succeeded in creating an ideological monster. It has acquired an autonomous life of its own, continuously gathering more momentum, running out of control, on a road to nowhere. This Frankenstein creation is running amok on the Earth, and as individuals we can seemingly do nothing but watch helplessly. It has become almost impervious to human will. Throughout the world, with terribly few exceptions, humanity rises no higher than being a mere agent of the capitalist vision of progress. Humanity is seen to have created a society where there are no real winners. Virtually everyone loses out to the 'system'.

In our everyday life we do not let this reality, nor any of the other madnesses observed by our extraterrestrial visitor, impinge too deeply upon our consciousness. We have to veil them from ourselves simply in order to be able to get on with life. However, in most cases, this necessary forgetfulness is not a matter of suppression, but of repression. Our awareness is censored not at a conscious, but at an unconscious level. Erich Fromm reasons that the most powerful underground force behind such repression is a fear of isolation and ostracism. This fear is so strong because it is rooted in a fear of identity loss. To be ostracized is to be disqualified from play, to be shut out of the game. It represents a loss of identity simply because competition in society's games frequently provides the only source of identity there is. When a player lives for the game, there can be nothing worse than being dropped from the team. The finite player represses their awareness of the unacceptable facts of our social reality through a fear of ostracism. It is the same powerful force as that which cuts the religious fundamentalist off from any awareness of the unacceptable facts of their spiritual reality. Fromm states the position perfectly in *Beyond the Chains of Illusion* :

> The individual must blind himself from seeing that which his group claims does not exist, or accept as truth that which the majority says is true, even if his own eyes could convince him that it is false. The herd is so vitally important for the individual that their views, beliefs, feelings, constitute reality for him, more so than what his senses and his reason tell him. Just as in the hypnotic state of dissociation the hypnotist's voice and words take the place of reality, so the social pattern constitutes reality for most people. What man considers true, real, sane, are the clichés accepted by his society, and much that does not fit in with these clichés is excluded from awareness, is unconscious. There is almost nothing a man will not believe—or repress—when he is threatened with the explicit or implicit threat of ostracism.[11]

The cohesion of the social landscape is contingent on conformity. The masses must be kept from seeing the reality behind the masquerade. The 'proper' functioning of society requires that all its finite games are actually played—and with all the appropriate seriousness. In many parts of the world this is ensured by dictate of authoritarian government. In the West, it is ensured with far greater subtlety through the sanctification of property. The possession of property represents security and power. It is title to property that we compete for in every finite game. It is our major preoccupation in life, just as it is the major preoccupation in the whole life of modern society. Indeed, today's burgeoning bureaucracies are largely devoted to the task of administering our entitlement to property. Concern with property absorbs an immeasurable amount of energy. The incessant scurry of human activity can be seen, ever producing wealth, ever competing for wealth, ever administering the allocation of wealth, ever consuming wealth.

Because our identity today is so bound up with our property, each society can be assured that its players will be prepared to support them in their power games with other societies. The power of the individual player—as measured in terms of their title to property—is invested in the power of the society to which they belong. Since a player's own power will automatically grow in parallel with the power of their society, it is consequently accepted that, as a society, we should always be striving for more power—power over other societies, who will in turn be striving for power over us. Our complicity in this game of escalatory power-seeking is guaranteed through the social cliché that is patriotism. In the passionate defence of our identity—as perceived in terms of our property—we blind ourselves to the reality that we are only defending an ideology, a collection of dressed-up fictions. The Cold War confrontation between the United States and the Soviet Union, for example, was not a confrontation between people, nor between politicians—politicians are just as much enslaved to the 'system' as the people are—but a confrontation between two fixed, sadly deluded, ideas about how society should be, two 'systems', feeding off each other, propagandizing their peoples, instilling fear in order to accrue more and more power to themselves, gathering momentum in subversive, tacit co-operation.

It is this momentum, the despotic, daemonic power of ideology, that is currently preventing the realization of a more authentic kind of progress. I believe that most people can vaguely perceive that our present situation is out of control, that they are being held hostage to anonymous ideological forces; the trouble is that for these same people the way things are is the only way things can be. Feeling utterly powerless, any sense of personal responsibility, any aspiration to idealism, is soon immobilized by the weight of what is called political realism. Thinking in the normal dichotomous manner, few can see beyond the traditional political poles of left and right. Anyone with any political ambition has to drop into one of the fixed categories of political thinking, all of which, bogged down in their own redundant dogma, stand for very much the same inadequate values. Tied up in both internal and external battles for power, the major political parties of the West have all lost sight of the fact that they are supposed to be serving the people, not their ideologies and not the 'system'.

The truth is that political parties invariably sacrifice their ideals to the practicalities of retaining power. Politicians are careerists rather than idealists. Within the framework of our current system, they have to be in order to survive. Votes are won by pandering to people's greed, not by appealing to their ideals. Consequently, modern politics has far more to do with kowtowing to short-term vested interests than any kind of long-term planning. The major political parties treat the electorate with utter contempt, and since we let them get away with it, their contempt seems well justified. The apathetic majority don't seem to care. As long as they can earn good money and don't have to pay too much in taxes, they will happily vote to perpetuate the status quo. Of course, being able to afford, for the moment, to insulate themselves from the environmental problems generated by our rapacious consumption, they do not have to, or want to care. They can easily persuade themselves that the price of complacency and apathy is one they can comfortably afford to pay.

But it is becoming increasingly more difficult to ignore our plight so casually. With all the media attention focused on the problems of the ozone hole and the greenhouse effect, most people are now abundantly aware of humanity's awesome responsibility toward the environment. Assaulted by a barrage of depressing, doom-laden statistics, most of us now understand that there are limits to the extent to which we can impose technology upon nature, limits to her raw material resources and limits to her capacity as a dustbin for all our waste and pollution. There is a growing awareness that we have entered a new era, a period where we *must* find a way of integrating our self-conceived technological landscape with the natural landscape. Many people are now beginning to realize that our mythology of progress is going to have to be rewritten. And the creation of this new narrative will require far more in the way of change than the cosmetic touches that are being worked on at the moment. The controls on CFC emissions and the introduction of lead-free petrol are encouraging developments, but represent no more than a slight alteration in the punctuation compared to the radical rewrite that is going to be required of us.

Technology has totally redefined our relationship with nature. By allowing us to incorporate nature into our finite games, it has encouraged us to claim the right to possess nature. But the truth is, of course, that nature cannot be possessed. We can certainly influence nature, but ultimately there is no way in which we can control it. The plain, simple, self-evident fact is that the finite relationship embodied in our present mythology of progress is unsustainable. To be authentic, any new mythology must embody a sustainable, infinite relationship with nature, recognizing that the well-being of the Earth and the well-being of its human inhabitants can never be separated. Any authentic mythology of progress must acknowledge the rights of the planet on behalf of all its living systems, present and future.

This infinite relationship with nature is my green rainbow. And chasing after it is our human destiny, the awesome challenge that will forever be presenting us with new problems and contradictions to resolve. Here too is the value which we need to ensure that technology will revert to being our tool, instead of growing to become our tyrant—for technological innovation will undoubtedly have a vital role to play in establishing an ecologically viable presence upon our planet. No one can deny the validity of this goal. No one can therefore deny that this is the goal toward which all authentic political activity must be aimed. It is axiomatic. The ultimate aim of every authentic political philosophy must be the creation of a sustainable and equitable society. Our political priority can be nothing less than the integrity of our planet. In terms of objective, politics *has* to be idealist. This means that every one of our traditional political ideologies—from conservatism and republicanism to socialism and liberalism—has to be considered thoroughly unauthentic. Their ideologies insist that there is no room for ideals in the modern world. Ideals simply don't pay their way. They cannot be entered on a balance-sheet. Such so-called realism is, of course, a sham. For political realism, I read surrender. For political idealism, I read hope. No one can honestly take any pride in claiming to be a realist.

Clearly, then, to talk about progress is to talk about political values and the nature of the political process. In the past, political philosophy has always been concerned with the relationship between the individual and the state. It is clear now that a third dimension must be added to enlarge the arena of political debate. With our new awareness, we now understand that the political process must acknowledge the rights of the planet. Political philosophy has to concern itself with the set of relationships between the individual, the state, and the Earth. The competing freedoms of individuals have now to be balanced against the freedom of nature. The conflicting rights of individuals have now to be balanced against the rights of the Earth. This is our great new political challenge.

It is my belief that this challenge will only be met through a strengthening of the bond between the individual and the Earth—at the expense of a weakening in the bond between the individual and the state. Only when power is devolved from the state—or that more anonymous entity, the 'system'—will people feel the sense of individual responsibility and participation that is so vital to a healthy culture. Today our culture is sick because the state exploits what is weak in human nature and stifles what is great. The state feeds on human insecurity by nurturing and encouraging our idolatry of consumerism. It does little to foster any independence of mind and develop the individual creative imagination. This is the spiritual bankruptcy of the materialist ideology which underlies both Western capitalism and the state capitalism of the East.

Humanity *is* capable of greatness, but it can far more easily succumb to the weaknesses of its own nature. By far the most serious of these is the tendency toward addiction—and not just physical addiction. Just as the mind can become addicted to drugs, so it can similarly become addicted to ideas. Ideas can take just as firm a hold on the human mind as chemicals can. The phenomenon can be witnessed all over the world, in every society, in every area of society. People place their ideological beliefs on an ivory ideological tower, refusing to question them, refusing to open their eyes to the reality that blatantly contradicts them, refusing to allow any contrary belief to intrude upon them, feeding their habit with a regular fix of religious or political cliché. Humanity will only be great, will only survive, through weaning itself away from its genocidal dependency on ideology. The enemy is neither communism nor capitalism, neither the Hard Left nor the New Right, neither the believers nor the non-believers, but ideology itself. We have got to start seeing beyond the familiar dichotomous categories to a deeper reality where there are no fixed solutions, a paradoxical reality which is home only to infinite play.

The paradigmatic shift in value that we need will only be brought about through individuals overcoming their need for illusions. Before we can free society from its ideological bondage, we must first free ourselves. The process must be catalysed at a personal level, at a cultural level; it cannot simply begin with some wonderful, idealistic plan for a new kind of society. That would be to fall straight back into finite play, playing to a pre-determined pattern. We have only to look at how the ideals of Christ were betrayed by the early Church, and more recently, how the ideas of Marx were betrayed by Marxism, to understand the dangers. History is littered with worthy ideals and ideas that have been distorted into disastrous ideologies. Once again, the only way we can truly proceed is through infinite play.

We live a paradox. Our ideologies have no external existence; they live and grow in the minds of ordinary people. We are their agent, and, at the very bottom line, their willing agent. Yet, collectively speaking, humanity seems largely unaware that it is a *willing* accomplice to its own enslavement. Humanity seems unaware that it is free to disengage itself from its ideologies. It believes that they are imposed from without when, in reality, it helps to impose them from within. The only way we can halt the runaway growth of our ideological systems is through becoming aware of our own freedom. We must help each other, by every appropriate means, to unveil that universal, rainbow-chasing, infinite player who lives within every one of us. As more and more people become infinite players, ideology will find itself with fewer and fewer agents to work through. Its shibboleths will become increasingly transparent to the common eye. Its power will gradually be eroded. This is humanity's way to freedom.

James Carse makes a vital distinction between society and culture. Whereas society is a finite game incorporating a large number of smaller finite games, culture is an infinite game—an open game in which everyone can participate. Whereas society is the realm of perceived necessity, culture is the realm of creative possibility. Deep down, there is a spiritual adventurer latent in us all. To participate in culture is to give free rein to that adventurous, visionary spirit, that poet in our heart, and, at the same time, to offer our experience to help unveil the poet that lives in the heart of every player of the game of life. In this way, culture allows vision to conceive vision. Culture is the vehicle by which vision can breed.

Infinite play is not for the faint-hearted. It involves us in the intensity of real life rather than the superficiality of charade. We look around and see people everywhere trying to avoid the action of the game of life. They shun the ball. If it comes their way, they will simply pass it on as quickly as possible. We see people barricading themselves against the demands of the infinite game, hiding from responsibility within the comfortable insularity of their boredom, untouched and unchallenged by the full potential of their humanity. We see people living in a fictitious, self-centred world of their own construction, secure and safe and comfortable. But the real world is not a safe and comfortable place, so by retreating into a manufactured world of security, they lose contact with the deeper, universal, infinite concerns of life. To compensate for the emptiness of such a superficial existence, we see people devoting all their energies to the objects of their lives—their possessions, their prestige, their persona. To play for the joy of playing is to have no need for this kind of self-aggrandizement. The present moment is simply too precious to be squandered in this way. We look for the action of the game. We pursue the ball and work with it creatively, bringing others into the play. Instead of seeking security by ever more closely defining the arena and fixing the rules of our games, we strive to extend their arena and experiment with the rules. Abandoning fixed roles, we disengage ourselves from the charade. We take off our masks and throw away the scripts. We play the game of life with our real self, spontaneously, lovingly, fully awake to its boundless possibility. We begin to understand that we are playing in a game that has no end. We understand that we can never claim a victory in the infinite game of life. The only prize is the miracle of our very participation. We start chasing rainbows, happy in the understanding that they will never be caught.

The joyfulness of infinite play, its laughter, lies in learning to start something we cannot finish.[12]

Proceeding in this way, by infinite play, it is not possible to describe a definitive mythology of progress. Our new mythology can only evolve as the play unfolds. What we can and must now talk about, however, is the first moves we should make in this infinite game of progress. Aware that a more authentic vision will require a shift from the abstract values of the market to the spiritual values of free individuals, I would like to suggest that we must begin with a radical reinterpretation of the meaning of education and democracy.

Understanding that we need to look to our children to solve the problems that ideological dogma is preventing us from solving today, we have to develop a system of education which considers the *whole* person, not just the intellect. We must encourage our children to seek value within their own potentiality, and thereby offer them an awareness of their own essential freedom. Only in that way can we expect them to respect the freedom of others—and the freedom of the Earth. We must also offer them a more authentic role in the political process of tomorrow's society. We must create for them a democracy which genuinely *is* a democracy, a democracy which is honest enough to reflect *their* values rather than those of ideological tradition. Only in that way, by offering them an effective sense of participation in society, can we expect them to make the effort to chase that green rainbow. We have to find ways of enchanting our children with the magic and mystery and sanctity of life on Earth. The hope must be that they will then be impassioned to assume the responsibility that we have so flagrantly scorned, the responsibility to maintain the integrity of *all* the diverse patterns of life on our planet, to create a sustainable human presence on the Earth. This must be the inspiration of any authentic mythology of progress.

8

Education as Enchantment

THE CRISIS of contemporary Western society is a spiritual crisis. Obliged to keep up with the hectic pace of life in the modern world, our sensitivities have been numbed to the wonder of existence. There is little in the technological landscape to mirror the essential dignity of our human being, little to stimulate a recognition of our greater potentialities. Despite all our time- and labour-saving gadgetry, the burden of living has never weighed heavier upon the human soul than it does in the hi-tech world of the late twentieth century. There is just so much that is required of us, so much pressure on us to win our finite games, so much to fear, seemingly, if we lose them. The 'system', like a huge parasite, sucks us dry of our creative vitality in order to fuel its own growth. Spiritually shrivelled, the tendency is for our religious instinct to be smeared out in a blur of banality. Experiencing few epiphanies of meaning, we are less and less touched by the spiritual yearnings of the human heart. The Western imagination has become alienated from the sacred mystery of life, estranged from its numinous source. Western culture has become spiritless, a culture of estrangement, a culture in which more and more people are living as strangers to their own being.

In moving toward a solution of both our social and our global problems, I suggest that we need to address one crucial issue above all others. How do we enchant our children with that 'delicate magic of life' referred to by D.H. Lawrence? How do we captivate their imagination in an age when the media has drowned the subtle under a tidal wave of triviality, when the creativity of art has been stolen by the stylishness of fashion? How do we fire our children with a desire for self-understanding when they are under persistent social pressure to devote their lives to the idol of consumerism? How do we develop in our children a spiritual sensitivity when they will likely be growing up in an artificial urban landscape, feeling no empathy with nature, their vision imprisoned by that sense of 'close-upness'? There are no easy answers. And there are some very daunting obstacles that will first need to be negotiated. For one, and perhaps most formidably of all, encouraging any such independence of spirit and imagination in our children would tread on all sorts of religious and political sensibilities.

State education, most particularly secondary education, belongs largely to the realm of finite play. Once children pass about their twelfth year, the great majority of their time at school is given over to the winning of paper qualifications. The school curriculum is given over almost entirely to examination subjects where the paramount aim is the acquisition of a designated set of facts. The objective is to have as large and as complete a collection of these facts as possible. Once they have been displayed in the final examination, they become largely superfluous. They have been converted into title. And most are quickly forgotten because they have made no real emotional impact. This kind of training obviously has a vital role to play in schools, but only in providing the foundation of literacy, numeracy and articulacy, and the framework of basic skills and knowledge upon which education must necessarily be built. Training and education are not the same thing. Training and education should *complement* each other. James Carse makes the distinction in this way:

> Education leads toward a continuing self-discovery; training leads toward a final self-definition.[1]

It is arguably useful to have some assessment of a child's various academic abilities, but there is no justifiable need to gear such a vast proportion of school time to just this one single aim. With so much emphasis placed upon academic success, the examination game has become highly divisive. It enforces the categorization of knowledge into well-defined, segregated disciplines, denying access to many important, less easily classified areas of learning. Philosophy and psychology, for example, rarely form even a small part of the school curriculum. The examination game naturally tends to categorize students too, as academic or non-academic, scientific or artistic, winners or losers. For the 'winners' school can often be traumatically competitive. For the 'losers' it will invariably be tedious and boring. It is seldom exciting—which is, of course, what education truly should be. Children should be graduating from school with their sense of wonder intact, sharpened rather than blunted, equipped and eager to carry on learning rather than glad to have put learning behind them.

Of course, a healthy proportion of children do find school exciting and do actually leave eager to learn more. The energy and dedication of large numbers of our teachers is such that many children do receive a pretty good education. Enthusiasm breeds enthusiasm. But I would suggest that success is usually achieved *in spite of* the system, and sometimes against quite appalling odds. If those odds are stacked too high, through a lack of resources and facilities, or perhaps through a basic lack of recognition and reward of the teacher's work and skills, that vital enthusiasm is liable to dry up. And the very children who are most in need of attention and encouragement will likely be the first to lose out. Far too many young people are leaving school without an interest in *anything* that they have been taught. Instead of being hooked on learning, they have become hooked on boredom. They enter into society with a paucity of vision which should be as shaming to us as it is to them. Their perception, possibly not unjustified when based on the evidence of their experience, is that education is not worth the effort, containing nothing that is worth caring about. They have been alienated because their individuality has never been properly respected. Much of what they have been taught is completely irrelevant to their needs and lacking in any kind of contextual meaning. The incalculable potential of their formative years has been squandered—irretrievably.

In such cases, of course, not all the responsibility, nor the blame, can be laid at the doors of our schools. Increasingly, children are isolated in the world of television, computer games and pop music, a world of sterility, devoid of any meaningful one-to-one interchange. Spending countless hours under the cataleptic entrance-

ment of the video image and the disco beat, many young people are
failing to develop the fundamental skills of literacy and articulacy
which represent the cornerstones of self-expression. Moreover, the
strange taboos of Western society prevent many parents from
establishing a healthy dialogue with their children. Young people
are unable to discuss the more sensitive, deeper concerns of life
because their mothers or fathers have somehow prevented themselves from being confidantes. On the surface, the family home
seems like a perfectly healthy, happy place, but there is little real
communication and understanding, little emotional intimacy, and
therefore little opportunity to develop any genuine emotional
sensitivity. In the most extreme cases, the lack of an ability to
understand and communicate inner feelings can surely be seen as
the prime cause of the despair which invariably lies behind acts of
vandalism and random violence. I see violence as a desperate, last-
gasp channel of self-expression when all other channels have been
blocked or have failed to be developed.

Violence is certainly not to be condoned, but neither is it to be
dismissed as evil. For many, particularly those who have been
brought up against a background of family and social oppression,
absorbing both the confusions of their parents and the contradictions of their society, violence represents a wretched cry for help
in a spiritual wilderness. It can invariably be traced back to a
frustration of the individual's essential humanity, a denial of their
emotional needs—the need to be loved and the need to be
respected—and a suppression of their natural potential—their
capacity *to* love, to be creative and to live a meaningful life.
Unfortunately, our system of education is not equipped, nor
mandated, to help such frustrated and alienated children pull
themselves out of the milieu of deprivation and hopelessness into
which they have been pushed by their family and social background. The teaching resources are not there. The sad fact is that
for those young people who have failed school—or, more properly,
have been failed *by* their school—there is often no further chance
of escape, and that all too frequently means that their own children
will remain locked into the same pattern of spiritual and emotional
insolvency.

In a society as incredibly complex as ours we carry a very heavy burden of responsibility toward our children's education. Having decided to bring a human being into today's confused and bewildering world, we should be prepared, both as parents and as members of society, to equip them as fully as possible for the demands of the difficult life adventure in which they will be obliged to participate. We owe this to them as an absolute moral right—a right which cannot simply be withdrawn when they enter into adulthood. It should be a lifelong right, for the demands of the game of life change with age. Each different stage of life opens up new areas of understanding that need to be explored. Education is not just for the young. Education should be freely accessible to *everyone* in society. There are some lessons that cannot be learnt until a certain level of maturity has been reached. For example, having little raw experience of life, young people are not capable of entering fully into the more profound and tenuous subjective areas of human understanding, nor is there any real need for them to, but the possibility should be there for learning later on in life, when the time is right, when the desire to embark on a journey of self-discovery eventually arises. Everyone should have the opportunity to acquire whatever navigational tools and expertise they need in life, at any stage of life.

Education, ideally, is the process by which we learn to affirm the values defined by the pattern of our human potentiality. Ultimately, education should be directed toward the satisfaction of our highest need, the realization of our highest value, Being. It should encourage us to live a life of intensity, a life of sensual participation. Education should provide us with the conceptual tools that we need in order to be able to build our own mythic bridge between the personal and the universal. Education should lead us into an ever richer discovery of our own creative genius, enabling us to perceive the extraordinary in the ordinary, to experience the whole that lives implicitly within every part. It should be an invocation of the sacred, an enchantment.

The power of this enchantment is very much dependent on our sensitivity to the magic of language, a sensitivity to the *poetry* of language, a receptiveness to the undertones and over-tones of meaning which give language its 'colour', an intuition for metaphor, a 'feel' for words which goes far beyond the possession of an extensive vocabulary. An empathy with poetic language is vital to our self-understanding because it gives us vivid access to the experiences of those who have gone before us. It also allows us to construct the metaphorical concepts that we need in order to comprehend our lives and our place in the world.

The more comprehensive and subtle our grasp of language— which must include a vital awareness of the limitations of language—the more finely tuned will be the tools that are our ideas, and the more realistic will be our grasp of reality. Since the way we think and behave is so thoroughly conditioned by our basic beliefs about what it means to be a human being, since we are so much a product of our own self-perception, the integrity of our Being-in-the-world is very closely related to the integrity of the ideas that we have collected in our mental tool-kit. It is these conceptual tools that we use when trying metaphorically to embrace our intuitive feelings, in order to make them accessible to rational thought. Effectively, our tool-kit represents our perso-nal mythology, our metaphysical map. If it is unauthentic or poorly defined, we can easily be led astray by ideas which hold out the appealing offer of security for our vulnerable ego. Without a well-defined map or a well-tuned set of conceptual tools, there is no foundation from which critical reason can operate.

The presence of reason is absolutely essential. Although it is right to want to exorcise reason as the all-pervading ghost of Cartesianism, we certainly do not want to banish it altogether. Too many people lack any logical base to their reasoning mind. Their thought has no guiding objectivity. And there are obvious sympt-oms of this. Most common, perhaps, are irrational prejudices against fellow human beings. Prejudice, whether it is racial, reli-gious, sexual or whatever, by serving to exclude, confers a perverted sense of identity and belonging through creating a membership of inclusion. More blatant still, and possibly more dangerous, com-pounding the problems created by prejudice, is the gullibility of so many to grossly over-simplistic religious and political dogmas.

It is nothing less than frightening to witness just how easy it is for ordinary, seemingly rational, basically well-meaning people to be hopelessly consumed by ideological doctrine. Although we all possess a common ground to our humanity, our personal conscious and unconscious are unique, born out of a unique genetic, societal and cultural cocktail. What we see, how it is interpreted and how it is acted upon, is conditioned by our own particular innate disposition and the specific events and ideas that we have experienced during our life. This is clearly a fundamental truth, yet bigotry remains commonplace in our society. People are seemingly unable to accept that the way they see the world is not the only way it can be seen, determined not to accept that, under different circumstances, the values to which they are so intolerant could easily have been their own.

Prejudices are fixed ideas or values which have taken root in the mind of an individual and function something like a polarizing filter on reality. Only ideas of the right polarity enter into consciousness. The transforming potential of all opposing ideas is blocked out. It is as a counter to prejudice that we find the most urgent value of literature within any culture. Not only does great literature offer us great ideas, it also attempts to rotate that polarizing filter on our awareness, allowing us to observe previously unrecognized aspects of our humanity. The fundamental aim of all great literature is to provide an insight into what reality looks like through a different filter of values from our own. By shifting the centre of our perception, we are offered a glimpse of the world through a different pair of eyes. In just the same way, but on a deeper metaphorical level, myth seeks to provide an insight into reality through a universal filter, an archetypal pair of human eyes.

Through stories—great epics, ancient myths, modern novels, tales of adventure—language binds our experience of nature to our experience of culture and serves to lift our spiritual awareness up by its bootstraps. The richer our experience of nature, which includes our experience of human nature, the richer will be our experience of culture, and, in turn, the richer will be our experience of nature, and so on. The key to authentic self-understanding lies in our ability to develop a resonance between these two modes of experience. And that is contingent in the first place upon our capacity to acquire—for ourselves—a sufficiently powerful set of metaphors with which to meet the world, to help us develop a full sense of the unplumbable depth of this whole improbable thing we call reality. The more profoundly we feel the mystery of our existence, the more we revel in the simple and precious joy of living.

In the Western world, almost everyone has the opportunity to improve their perspective on reality. By climbing on the shoulders of our great mythic explorers—the countless scientific, artistic and spiritual visionaries that have ventured out before us—we should all be able to see further than ever before. But most of us, it seems, don't bother. It is too much effort. Having their awareness channelled through well-established conduits of socially accep- table thought, people prefer to clutch hold of ideological slogans rather than seek the inevitably more uncomfortable and compli- cated truth. Equipped only with crude and feeble conceptual tools, they can only fashion a very crude and feeble world-view, a view full of simplistic explanations. For many people, that essential resonance between experience and thought has never developed because their navigational tool-kit of metaphors has simply been too impoverished to initiate the bootstrap. They have been filled with facts, taught techniques, inculcated with clichés, but have never been given the opportunity to develop the inner strength with which to lift their self-understanding up by its own bootstraps. This is not something that can be done *for* anyone. We have to do it for ourselves. Meaning cannot be taught. It has to be constructed internally and experienced, individually. A beautiful metaphor is provided by this little tale from Idries Shah's collection of Sufi teaching, *Thinkers of the East*:

Someone said to Bahaudin Naqshband:
'You relate stories, but you do not tell us how to understand them.'
He said:
'How would you like it if the man from whom you bought

fruit consumed it before your eyes, leaving you only the skin?"[2]

This story can be seen to function as a kind of meta-story, a story offering a meaning which can only be experienced within the listener, about the fact that the meaning of such stories can only be experienced within the listener. I therefore don't intend to explain. The voice of metaphor simply speaks so much more strongly. Metaphor is capable of carrying a meaning that lies beyond the realm of explanation. It can penetrate our awareness in a way that explanation simply cannot. Here is another short tale from *Thinkers of the East:*

> A certain important man of learning said to a Sufi:
> 'Why do you Sufis always use analogies? Such forms are good enough for the ignorant, but you can speak clearly to people of sense.'
> The Sufi said:
> 'Experience shows, alas, that it is not a matter of the ignorant and the wise. It is a matter that those who are most in need of a certain understanding, or even a certain part of understanding, are always the least able to accept it without an analogy. Tell them directly and they will prevent themselves perceiving its truth.'[3]

Metaphor can do what explanation cannot possibly do; instead of offering rational knowledge, it steers us into original thought and feeling, toward a deeper, intuitive understanding of ourselves and our world. But, paradoxically, that understanding is not to be found within the metaphor itself. As so exquisitely expressed by James Carse:

> It is not the role of metaphor to draw our sight to what is there, but to draw our vision toward what is not there and, indeed, cannot be anywhere. Metaphor is horizontal, reminding us that it is one's vision that is limited, and not what one is viewing.[4]

The world cannot be contained by the word. Sensation cannot be touched by semiotics. Rational analysis cannot compete with the *experience* of being touched by a great myth, story or poem. Playing with language in an infinite way, such narratives do not seek to explain. Instead, they offer insight. They speak to the unconscious depths of our understanding. They evoke empathy. They invite us to respond with our own original feelings. In this way, to borrow Carse's phrase, the 'reading of the poetry is itself poetry'.[5] The experience of an infinite narrative offers us the vision to see beyond our personal perspective, opening our vision to a universal realm of human experience and values, not by providing us with answers, but by raising questions which challenge us to confront the unbridgeable gap between experience and what language can say about experience.

Explanation sets the need for further inquiry aside; narrative invites us to rethink what we thought we knew.[6]

Sadly, though, in our conservative and complacent society, few people are open to rethinking what they think they know. With their personal mythology long ago petrified into an ideology, most have indeed set aside the need for further inquiry. Having been trained rather than educated, having savoured nothing but explanation—what I would call pseudo-scientific and pseudo-religious explanation—they have been deprived of the delicate taste of the fruit. They have been left with the skin, and no appetite for anything but the superficial. This is surely why so many fall such easy prey to today's junk culture. Its products are instantly consumable—and consuming, consuming of individuality and creativity and spirituality. With their capacity for wonder withered away, with their spirit left untouched by the ideals of our great myths, the aspirations of the larger part of Western society are driven by the brainwashing techniques of a multi-billion pound consumer industry. There is now so much that can be passively consumed that we seem to be losing the ability to use our imagination actively to originate entertainment for ourselves. Worse still, we even seem to be losing the ability to enjoy ourselves.

This hints at possibly the saddest contradiction in Western society: the near ubiquity of boredom in spite of all the dazzling technology which is used to try to keep us entertained in our leisure time. The plain fact is, though, that we are generally not very well practised in the *art* of enjoyment. Enjoyment is a celebration of our humanity. It is the response of an active, concentrated mind taking satisfaction in the productive use of its talents. The key word is concentration, and this, indeed, is the very skill which is so difficult to develop within a culture where every second seems to count so much. By attempting to extract as much as we can out of each moment, our attention becomes too unfocused, life slips by unlived, experience slips by unenjoyed. We eat food without tasting its flavour; we listen to music without hearing its content; we look at the world without seeing its shapes and colours; we read words without absorbing their meaning. The mind is always somewhere else.

Very few of us can avoid the pressures imposed on us by our complex social system. It is the price we pay for all the incredible luxuries and conveniences that we now take for granted. The trouble is that it has all become *too* convenient, *too* comfortable. As we are tempted to invest more and more of our energy in outer as opposed to inner security, as we are increasingly tempted to seek out things rather than experience, so we feel less and less desire to escape our material incarceration and explore the fantastic natural and imaginative landscapes that can put us in touch with the enigmatic mystery of our own being. The conceited march of technological progress has stolen our awareness from the realm of the inexplicable and steered it toward the explicable, collapsing the spiritual dimension of life, squeezing out the more profound and subtle aspects of existence, diminishing the astonishment which we should always be feeling before the sacred mystery of life. It is not often that we can genuinely enter into the full wonder of our being because the demands of the immediate moment are always perceived to be too great. We rarely appreciate the sanctity of the Earth because we rarely have time and cause to think about it amid the all-consuming exigencies of modern daily life. There is either too much else to think about, or no energy left to think about anything. With so many disenchanting preoccupations in our lives, there seems to be little opportunity to take out that regular slot of time which is needed for our self, to feel the connection between our inner and outer worlds, to search, in this relationship, for our spirituality. It is difficult to concentrate long enough to hear our own original voice above the general cacophany of social clichés.

Acknowledging that our mind is intrinsically lazy, only too ready, so it often seems, to idle away in a nice, undemanding gear, we need symbols, reminders, stimuli to our self-enchantment, to focus our attention, to kick our consciousness into a higher drive, on to a greater plane of spiritual awareness. This is why it is so important that we each define for ourselves a sacred place or object or activity, a totem, which, for some special reason unique to us as an individual, speaks to our soul, lifting our consciousness above the crushing triviality of modern everydayness. Many find their sacred totem in their church. Others, like me, may find it in the lonely wilderness of the mountains. Some may find it in painting or music, or perhaps a physical discipline like, in my case, running, or *T'ai Chi* or dance. It does not matter. Sacredness is not an absolute. A place or an activity is not made universally sacred through a ceremony of consecration administered by an appointed religious authority. A certain totem is made sacred to us, personally, because it communicates the spiritual mystery to us, because it leads us into a recognition, both humbling and ennobling, of our innate human dignity, because it has, over time, evolved to be our own particular mantra of authenticity.

Sacred time, time for meditation, together with time for personal creativity, time to pursue our own particular choice of ecstasy, is never selfish time. This is time for concentrating our awareness, for finding Silence, for bringing our mind and body together in psychosomatic harmony. This is time when we make a withdrawal from the interest earned on the accumulated deposits of our mental and physical training, when we cash in the bonds of years of effort and discipline, when we pursue enjoyment through the creative use of our particular skills or fitness or understanding. And we find that this productive activity is an educative process in its own right. To follow our bliss is always to enrich it, to realize our natural genius with ever greater flair and diversity. Whether it is to seek an inner quietness or to follow an outer passion, this is time to fill the soul with gumption, to enchant ourselves with the game of life.

Born with a rational mind and the capacity for free, independent thought, we enter life with an innate sense of estrangement. But it is this very estrangement which generates the question to which our life, if lived authentically, represents the answer. By living in open acknowledgement of our paradoxical separateness from nature, we fully fuel our desire to create meaning, which authentically finds its ultimate expression through love. Enchantment is the process by which estrangement is turned into love. The desire to care, to understand, to respect, to respond—to *love*—flows out of enchantment. And that is what education, authentic education, is truly all about—developing our sensitivity to enchantment, our fascination with the wonder of all that is. And that, too, is what life is truly all about. Lived authentically, life is a journey of continuing self-discovery, of becoming ever more wholly human. There is no end. The only end is the journey itself. The ultimate mystery is never to be solved, but its paradoxicality is to be entered into ever more fully.

It is through culture that our humanity becomes manifest. As James Carse succinctly expresses it, 'One cannot be human by oneself.'[7] The journey of self-discovery cannot be undertaken in isolation from either the world of nature or the world of the collective human imagination. Our humanity is rooted in our language. Although self-understanding is grounded in experience, which is ultimately untouchable by the word, we still need language in order to be able to interpret that experience. It is our culture—the communicated experiences of those that have gone before us—that allows us to discover who we are, and it is our *participation* in our culture—the communication of our own experiences—that creates meaning by contributing to the self-understanding of those who will hopefully follow us. Just as we use the stories of others to help us understand our own experiences, so we in turn hope that others will be helped to understand some of their experiences in terms of our stories. It is in just this way that love and language bind us together as a people. It is love and language that allow us to touch the inner world of another from within our own inner world, to engender care *with* care, to create understanding *with* understanding, to inspire respect *with* respect, to promote responsibility *with* responsibility, to raise laughter *with* laughter, to kindle joy *with* joy. Enchantment is that vital resonance between our experience of nature and our experience of culture, the dialectic nucleus of education.

Education as enchantment joins people culturally through a network of communication, introducing individuals to the natural and conceptual experience which they need in order to fulfil their creative and spiritual potentialities. Sadly, though, many of our schools are more a home to disenchantment than to enchantment. Most particularly, with their ridiculous emphasis on 'facts', schools either just ignore those realms of inquiry which are home to mystery, or seek to neutralize their vitality with spuriously simple explanation. They often succeed in snuffing out the fire of the creative imagination rather than fuelling it. In fact, it could well be argued that schools actually serve as society's means of *preventing* children from being exposed to education, for fear that too many might come to see through the illusions which maintain the status quo. Indeed, perhaps this is why compulsory free schooling stops at just the age when young people begin to become self-motivating!

For the children of our aboriginal peoples, education was mediated through myths and rituals and an extensive participation in daily cultural life. Their language and mythology, having evolved in intimate relation to the landscape in which they lived, provided a clear sense of place and belonging and an understanding of the natural patterns of life. Aboriginal people lived in great respect of these patterns. They lived in harmony with them. They had to in order to survive. Their life was a life of enchantment. This is why their whole living landscape was sacred. How different is our life today. It is only a rare few who manage to live a life of enchantment amidst the staggering inhuman complexity of modern society. It is only a rare few who genuinely, affectively, feel a reverence for the natural world. Again, our most urgent need today is to acknowledge and address the problem of how, in a culture of increasingly subversive superficiality, we enchant our children with a sense of the sacred.

Our being-in-the-world has no foundation accessible to human rationality. Sooner or later, one way or another, this ultimate intransigence of the human condition has to be set as the keystone of the mythology which is offered for our children's education. Because it represents an open narrative, expurgated of explana-tion, there is no danger of it being hijacked by ideology. It will point to the sublimity of our sensuous presence in the world. It will offer each child a mirror in which they will be able to see themselves as the living answer to their own existential question. It will offer that mirror as the only vision of truth that they can have. Paradoxically, the more deeply we experience the meaning of our own inexplic-ability, the greater is our awareness of our own genius. The more profoundly we understand what it is that lies beyond under-standing, the richer is our enchantment, and the more intensely felt our reverence for the Earth whose children we all are.

Although our aboriginal peoples were nomadic, their imag-ination had a distinct home in the lands that they roamed. Their psyche was firmly and healthily rooted in the Earth. Today, the Western imagination has no such roots. We have splendid homes, full of all sorts of incredible material possessions, but our imagination is essentially homeless. It has no spiritual sense of place. This strange homelessness of the modern imagination can only be overcome, at both an individual and a social level, by acknowledging the real nature of human freedom, recognizing that, in our psyche, we still remain essentially nomadic. As James Carse says:

> Human freedom is not a freedom over nature; it is the freedom to be natural, that is, to answer to the spontaneity of nature with our own spontaneity.[8]

Nature has no voice of her own. She is the unseeing mirror in which we experience the reflection of our own genius. Our own divinity. She provides the estranging silence which actually creates the very possibility of enchantment. This points to the multi-faceted paradox, the crucible of inexplicability, which Western civilization has utterly failed to embrace and which sits at the heart of our contemporary crisis in meaning: the complementarity of our relationship to nature—that we are both estranged by the total silence of nature and quite undivided against nature in its totality. The silence of nature is total because we exist *as* nature, inseparable from its totality. We cannot look at nature for the same reason that we cannot look into our own eye.

The global problems that we face today are not going to be solved within a generation. Our most urgent social responsibility is to educate our children so that they might recover the mythic sensitivity and intuitive vision that we, as a people, haven't had the courage to search for ourselves. It is out of the seed of our infinite play with our children today that the strength of our collective vision will grow tomorrow. We must light the creative imagination of our children with the touch of our own creative imagination, not by imposing upon them our own particular pattern of thought, but by offering them the opportunity to develop their own thought and their own feelings, grounded in first-hand experience rather than second-hand opinions. Perhaps foremost, this experience should include the experience of infinite narratives, myths, those great stories and legends from which the timeless truths of the human spirit speak out to us, engaging our intuition with their spellbinding voice, opening the imagination to wonder and the deeper patterns of our instinctive potentiality. In these treasuries of archetypal wisdom are to be found the metaphors which allow us to commune authentically with the mystery of our own originality. These allegorical symbols, the unconscious product of the boiling down of universal experience to its spiritual essence, are among the most important tools that we can offer our children, equipping them to be able to discover for themselves the true nature of human freedom, encouraging them to experience their kinship with the Earth as both the physical and the spiritual body of their most profound existence. This way, we might hope that they will find the courage to live the exemplary life, the life of enchantment, that we, through our infidelity to the sacredness of the Earth, have so tragically renounced.

9

Evolution as Involution

O UR INTERPRETATION of scientific and religious mythology has revealed that both systems of thought, beneath their dualistic veneers, conceal the same fundamental implication. They each tell us that, just as surely as the personal is embodied within the universal, so the universal is embodied within the personal. The personal and the universal, subject and object, are recursively engaged in a mystical symbiosis. This logically inassimilable reality has been something of a reverberant theme for us, resounding, either explicitly or implicitly, through the course of much of our discussion. I would like to suggest, then, that we have here the axis around which to write a new story of the Earth. Here is the narrative context for a new creation myth, a new story of origins, to supersede the impoverished stories that continue to be peddled by our two great cultural institutions.

It is in this central aspect, the inadequacy of their stories of creation, that we presently find both traditional science and traditional religion being brought most conspicuously into disrepute. However, so deep run the conduits of institutionalized thought that no other story ever gets an airing in our media-tuned, media-directed, media-controlled culture. You either have the standard scientific theory of evolution or the stultified creation myth of Genesis. Once again, we seem to be dictated to by the either/or mentality of our culture. Life was *either* created at the whim of an almighty God *or* it has evolved as the fortuitous result of blind chance. Just two alternatives. The result is that, faced with having to choose between these intuitively unsatisfying, irreconcilable extremes, most people have got caught in a kind of metaphysical no man's land, having very little idea quite what story they believe and, with confusion breeding apathy, no real interest in finding out. And without a unifying story of origins, there is no narrational bond to join people together in spirit. This is how we find ourselves today. There is no cultural communion, no cultural community, no cultural commitment.

An origin is a source, a centre, the point to which all other points are referred. Without a mathematical origin, for example, we can have no way of defining a point in space. Similarly, without an existential origin, we can have no way of defining the point of our own being. Without an originating story, we can have no existential frame of reference, and no genuine originality. As I suggested in the Introduction, the Western consciousness is presently in its spiritual adolescence, groping into maturity, struggling to accept the demanding terms of adulthood. We are living today in a period of transition. Having moved from a revelatory story to a rational story, we are now coming to realize that the irrational cannot be written out of the plot so easily. Slowly, our collective imagination is beginning to embrace the idea that a rational story cannot, in principle, be the full story.

By signifying a capricious, inauspicious origin amidst the random collisions and interactions of organic molecules in a chemical soup, the modern scientific theory of evolution has effectively eliminated any sense of the immanence of the divine within the life process. Also, rightly or wrongly, by presenting our being-in-the-world as purely accidental, traditional science cannot help but confer an ignominious sense of futility upon human proceedings. Being is stripped of its sacred quality. It becomes an improbable, adventitious intrusion into what is seen to be a ruthlessly objective universe. However, what is so desperately wrong with the scientific story of origins is not so much the idea that we have evolved out of a volatile primordial soup of chemicals—which, in objective terms, seems clearly to represent the basic gist of things—but that, in the earliest life of the universe, there is no acknowledgement of any kind of subjective reality. The ideology of science dictates that the reality of the universe is fully and solely contained by its physical-material dimensionality, and therefore fully explicable in terms of the behaviour of its constituent objective parts.

This dogma, which in this book we've been consistently seeking to subvert, is what has undermined our sense of our own divine relationship to the world. We no longer see infinity in a grain of sand. We see merely a crystalline lattice of silicon dioxide. The point here is not that the scientific experience has replaced the spiritual experience—because, properly, our scientific perception should enhance our spiritual perception—but that we have been brainwashed into thinking that the scientific experience is the total experience. We have acquired a habit of thought which filters out any deeper, intuitive, religious experience. Most of all, it is this very feeling for reality in its numinous depths that must be recaptured by our new creation story. For without it, not only will the organic functioning of the Earth's life systems disintegrate, but our very humanity will disintegrate too. To assault the Earth is to assault our own being. And this is just what we are surely doing.

What we need is a more intuitively appealing narrative of origins, one which cross-fertilizes the two rather narrow visions of creation appropriated for us by our scientific and religious communities. We need a story which embodies wholeness as its leitmotif, affirming that, in the words of Kathleen Raine, 'the outer world is within us; and our inner world is that very world we see, hear, feel, touch'.[1] This new mythology will certainly accept the fact of evolution, but it will not accept that it obeys the mechanical rationality of the 'blind watchmaker'. It will see life on Earth, as the evolutionary product not of a material but of a mental process, a process with two manifest levels, psychic and physical, each evolving toward ever more intimate mutual involvement.

This very much represents the kind of vision which resounds through the inspirational prose of *The Dream of the Earth*. Owing a considerable debt to the ground-breaking, if slightly flawed, work of Pierre Teilhard de Chardin, Thomas Berry's work is ground-breaking in its own right by perfectly capturing the nature of our ecological problems, and by laying at the door of each one of us the nature of the task that lies ahead if we are going to progress toward any kind of solution. He has described himself as an 'ecologian'. It is a delightfully apt epithet. To read the words that have flowed from Berry's pen is to meet a man who is passionately, religiously, in love with the Earth. He writes:

> If we were truly moved by the beauty of the world about us, we would honour the earth in a profound way. We would understand immediately and turn away with a certain horror from all those activities that violate the integrity of the planet.[2]

The reason we are not sufficiently moved by the beauty of the world is that we don't feel sensuously involved in it. And this returns us once more to the story of creation. We feel no intimacy with our world because our rational story tells us that we have evolved *out* of it, as a bubble of subjective consciousness, a world of our own, separate from the natural world from which we were born. But, as we are now very well aware, this story is seriously flawed. There is no detachment. There is no separation. Our conscious mind is contiguous with our unconscious mind which is contiguous with our body which is contiguous with our world. Our evolution is the evolution of this whole system of being. Our evolution is the evolution of what we earlier called Mind, the mind-at-large in which each individual human mind is a co-evolving sub-mind. It is in this context that Thomas Berry sees our human being activating the most profound value of the universe, 'its capacity to reflect on and celebrate itself in conscious self-awareness'.[3] In another beautiful expression, he says that 'we emerge into being from within the earth process and enable the universe to come to itself in a special mode of psychic intimacy'.[4] It is our brain which makes this intimacy possible. With more than ten *trillion* nerve interconnections packed into a mass of tissue weighing no more than three pounds, the human brain is where the psychic and physical levels of the evolutionary process are most finely cross-mapped, where the spiritual and material realities of the universe are engaged in their most intimate mutual involvement. In the labyrinthine connectivities of the brain, we find subject knotted with object in the most unfathomably intorted and awesomely mysterious entanglement in the entire known universe.

Far more mysterious than outer space, the inner space of the human mind confronts science with questions that it would actually rather not have to answer—indeed questions that it tries very hard not to have to answer. As I suggested earlier (see Chapter 1), through its determination not to ask itself the really awkward questions, traditional science discloses an irresponsible, conceited kind of complacency. Its demystification of matters that it has no right to demystify has served to lend a material rather than a spiritual shape to our deepest sense of ourselves. The central task of this chapter, then, is to restore some of the mystery that has been stolen away by the scientific story of creation. Only in this way, by recovering its own primordial mystique, can the Western consciousness hope to recharge its originality with meaning, and so recharge our lives with an authentic sense of being.

It must be stated at the outset that evolution in itself is not in question here. In the face of a wealth of scientific evidence, evolution has to be accepted as a fact. What *is* in question is the nature of the process of evolution. I believe this question is of vital importance because the Western vision of the world has become heavily influenced, and damaged, by the surreptitiously pervasive dogma that takes our consciousness to be nothing more than an accidental intrusion into a thoroughly material universe. Unfortunately, despite the fact that this belief rests upon outrageously thin foundations, the scientific story is not that easily remystified—at least, not in the rigorous way that science demands. It is, of course, not without very good reason that this explanatory narrative has such a strong hold on Western thought. The trouble is that, on the surface, it is so eminently believable. Rationally, it is completely unassailable. Intellectually, it is dangerously beguiling. For years, I lived under its seductive spell. In the Prologue I described the intuitive experience which broke that spell, where I crossed some kind of spiritual Rubicon. I now want to present the line of reasoning which I subsequently developed in order to rationalize that spontaneous experience of disbelief.

First, though, I have to admit to experiencing difficulties with this task. Even now, the analytical side of my mind remains partially bewitched by the logical elegance and sweep of the scientific theory. I can still remember the smell of the intellectual power that I felt over people who wanted to see supernatural forces at work in the evolutionary process. For me, at that time, such beliefs were taken to betray a certain intellectual feeble-mindedness, indicative of a failure to see into the stochastic roots of creativity and the staggering numerical timescale of evolution. When I now take up my counter-position, I can see the same glint of power in the eyes of the rationalist that I once must have had sparkling in my own. It is an uncomfortable position for me. Having been on the other side, I appreciate only too well the weight and might of the heavy intellectual artillery that bears down upon me. The great difficulty is that my argument with the scientific story cannot be quantified. The points I want to make are inherently qualitative and fail, abjectly, to satisfy the criteria of scientific proof. I intend to suggest that the various narratives which go to make up the main plot of the scientific story are, in some way, mutually inconsistent, yet these sub-plots are so incredibly complex, and so inadequately understood, that a conclusive proof is quite unattainable.

The discussion which follows has been specifically written with the scientifically inclined reader in mind, for it is those of us with a work-hardened scientific mind-set who are perhaps most vulnerable to the intellectual lure of scientific explanation. Scientific scepticism can only be met with scientific argument. The deepest intuitive conviction will add nothing to the open forum of rational converse that is needed in order to loosen the grip of dogma upon the scientific consciousness. My main aim is to raise the temperature of this presently rather lukewarm debate. I hope I have provided enough background information to allow the main body of the argument to be followed by anyone, but some of the detail may well be difficult for those readers who are entering into unfamiliar territory. Although the more spiritually inclined reader will probably want little persuasion that the scientific story is not the whole story, I still believe that the discussion will be of value because any understanding of the scientific perception of things can only enhance the quality of our spiritual perception. Many parts of the scientific story are quite awe-inspiring. Their narrative beauty and power demands our appreciation in spite of any reservations we might have about the completeness and consistency of the whole.

The natural place to begin our discussion is with deoxyribonucleic acid, or DNA, the very centrepiece of the modern scientific story. DNA is the famous molecular blueprint which sits in the nucleus of every living cell. It consists of two long chains which are twisted around each other in a spiral called a double helix, each chain consisting of a sequential string of fairly simple molecular structures called bases, tied together like a string of beads. These chains are incredibly long. A typical DNA string will contain millions of beads. Each DNA molecule in the cell nucleus is tightly wound around itself and protected by proteins in a package called a chromosome. In the nucleus of human cells there are forty-six such chromosomal clumps of DNA, each housing many thousands of genes—the functional units into which each DNA string is semantically divided. The beads come in just four different kinds: adenine and thymine, guanine and cytosine. The pairings are important. In their helical intertwining, only adenine can sit opposite and link with thymine, and only guanine with cytosine. This means that one chain of bases in the spiral completely defines the other by default. It is this property which makes it possible for the DNA molecule to replicate itself.

The DNA chain can best be thought of as a vast program for a chemical computer, with each gene representing a sub-routine or sub-program. Very much like a piece of computer software, the DNA chain contains both data and instructions, all written in a special code. Although our understanding of the instruction set is still very vague, the chemical code itself was cracked back in the 1960s, and this has enabled us to interpret a good part of DNA's program, most notably its huge protein database. Here, with a beautifully simple ingenuity, each set of three bases, called a codon, is mapped onto a certain symbol of a molecular alphabet. Since each location in a codon triplet can be occupied by any one of four different bases, there are sixty-four possible combinations in total; however, there are actually only twenty letters in the alphabet, plus punctuation symbols, some being mapped on to just one codon, others on to as many as six. Knowing this set of one-to-one correspondences between codons and the letters of the molecular alphabet means that the DNA can be read as a text. This is what happens in the cell. In fact each letter corresponds to a specific amino acid, one of just twenty quite simple organic molecules which represent the fundamental building blocks in the construction-kit of proteins, which are themselves the most important building blocks in the construction-kit of life. The DNA's text tells the cell which blocks to use, and the sequence in which they are to be put together, in order to construct the body's essential proteins.

These giant molecules, providing the structural framework and the active processing agents of all living cells, are almost entirely composed of amino acids, hundreds in number. Depending on how the building blocks are arranged, proteins fall into two broad categories: fibrous and globular. As implied in the name, fibrous proteins possess straight chains and generally have a structural function; globular proteins possess tightly, irregularly coiled chains and have a more dynamic role to play, their function being largely determined by their three-dimensional structure. The most important globular proteins are enzymes. These are what appear to be custom-built molecules, with an active site which has a very highly specific shape, like a precision-engineered, three-dimensional key, designed to fit a particular chemical lock within the cell exactly. Almost every one of the thousands of potential reaction pathways in the molecular soup of the cytoplasm—the volume of the cell surrounding the nucleus—is catalysed, or opened, by just one specialized enzymic key. The shape of this key has to be perfectly tailored to the shape of the reactants, to the extent that the slightest mismatch will upset their complementarity and keep the reaction pathway closed. Even more remarkably, the three-dimensional structures of proteins, dependent on relatively weak hydrogen bonding, are very delicate. In fact, they are inherently unstable. The cell must maintain a very finely regulated acid-alkaline balance and temperature to preserve the integrity of its proteins. Life is completely dependent on the existence of these proteins, but their own existence, in turn, is completely dependent on life!

Although the covalent bonds between bases along the length of a single chain of DNA are quite strong, the complementary bases on the two twisted chains, the links between adenine and thymine, and guanine and cytosine, are hydrogen-bonded and are consequently quite weak. This means that the DNA molecule can be separated like a zip, a portion of its message transcribed onto another molecule, called messenger RNA, and then zipped back again. The messenger molecule can then return to the cytoplasm and get its coded message translated. All sorts of perfectly adapted molecules functioning as a team, each with a well-defined task in the manner of a factory production-line, combine to read the codon, pick up the associated amino acid, and build it, molecule by molecule, into the required protein, finally stopping when the 'full stop' codon is reached.

Needless to say, the actual biochemistry of all this is stupendously more complicated than it is possible to indicate here. The single cell of a typical bacterium, for example, synthesizes possibly 3,000 or more different protein molecules simultaneously, all manufactured at exactly the demanded rates to maintain the concentrations needed for the cell's required function. For me, this paints a truly awesome picture: perfectly counterpoised feedback controls and feedforward cascades, self-regulating auto-catalytic loops, every chemical reaction in a frenzied kind of balance with every other, all immaculately interlocked as one co-operative, integral unit.

The DNA molecule plays the only passive role to be found amidst all this highly orchestrated chemical chaos. It sits in the nucleus like a huge manual of instructions, messengers continually commuting back and forth from the cytoplasm, transcribing individual pages of text and turning the message over for translation. Of course the DNA not only contains the instructions for all the protein assembly work, but also contains all the higher-level instructions determining which particular proteins each cell is to produce, telling it, in effect, which kind of cell it is going to be. It does this by controlling the production of repressor molecules which bind the unwanted pages of the manual together so that they cannot be read—in other words sealing the zip of the double helix.

In this way, the messenger molecules, many thousands of which are accessing the manual at any one time, are only able to read from the right pages. It will be no surprise that exactly how all this is encoded within the DNA molecule is still very much a mystery.

My first introduction to the complexity of this world of molecular industry came about through reading, albeit casually, A.L. Lehninger's definitive *Biochemistry*. It was an overwhelming intellectual experience, no less fascinating than my first close encounter with the weirdness of quantum physics. To conceive of the full complexity of just a single cell is impossible enough; if we extrapolate from that to the one hundred billion or so cells that go to make up the human organism, with the activity of each cell perfectly integrated with the activity of every other cell so that they all function together as one whole—to conceive of that order of complexity is impossible to the point of being apoplectic! However, this is far from the end of the story. The fact is, of course, that our DNA is not only the software which oversees the functioning of all our organic hardware, it is also the software which builds all that hardware in the first place! Morphogenesis is the name given to the process by which a single fertilized cell divides and multiplies to form a fantastically complex living organism. By way of logical mechanisms which remain almost entirely unfathomed, every cell ends up in exactly the right place, connected up in exactly the right configuration, specialized to do exactly the right job. Out of the DNA's inherently one-dimensional message is born a physical form with a multi-dimensional logical coherence.

And the story, of course, goes on yet further. By far the most incredible logical mechanisms of all are to be found in our brain. The logic circuits here are constructed out of billions of nerve cells called neurons, each of which consists of a cell body with a delicate network of filaments, like a root system, branching out from either end: the dendrites and the axons. Roughly speaking, the dendrites gather signals and the axons distribute them. There is actually a certain functional similarity between neurons and the transistors out of which digital computers are constructed. Although structurally they are worlds apart, especially in that the neuron has a thousand or more inputs to the transistor's meagre two, the basic operating principle is much the same. When the sum total of electrical input at the dendrites, positive balanced against negative, exceeds a certain threshold, the neuron will electrically fire, transmitting a potential to the axons, which then feed this signal, across tiny gaps called synapses, into the dendrites of many other neurons, some of which will consequently be tipped over their firing threshold in turn. And so the process continues, incessantly, billions of firings a second underpinning your reading of these words at this very moment!

Although it is convenient to talk in terms of electric currents passing around the circuits of neurons in the brain, it is important to point out that these currents are not transmitted by electrons, as in digital computers, but by charged ions, principally sodium and potassium, and their flow is subject to all manner of different molecular controls. The electrical logic of the brain is really just another dimension of its molecular logic. The two interact so closely and are so mixed up with one another that they cannot be separated. They are different aspects of the one logical calculus. Charged ions and molecules are the vehicles for the transmission of information around the neuronal circuits of the brain. However, it would be very wrong to think of individual molecules as carrying discrete bits of useful information. The software symbol level must be much higher, involving large ensembles of molecules locked into dynamic but stable patterns. We know as little about this symbolic architecture as we know about the hardware architecture of the brain. How logical programs are encoded within the patternings of the neuron network is an almost total mystery.

Like a digital computer, the brain, at the very bottom level, appears to operate by way of a very precise logical calculus. However, the end result turns out to be very unlike a digital computer. The big difference is in the fantastic multiplicity of connections that each neuron can make. The brain's logic is fuzzy. And its architecture is fuzzy. Whereas in the digital world there is a fairly clear demarcation between software and hardware, within the brain there is no real way of saying where the hardware ends and the software begins. The molecular software actually constructs and modifies the hardware. Although it is easy to see how this possibility could make for tremendous programming power, it is not so easy to see the mechanisms which prevent the whole system from sliding into anarchy. With so much fuzziness, with so much freedom of information, how is it that order rules over chaos? How does our unique sense of self emerge out of the brain's inordinately complex flux of molecular information? These are profoundly perplexing questions to which science has no answer.

The most obvious differences between brains and computers are to be seen in their disparate functional abilities. Our brain is quite useless at the number-crunching at which digital computers excel. We even have trouble remembering a ten-digit telephone number. Yet, our brain is quite superlative when it comes to real-life problems like recognizing patterns and extracting meaning from a field of data, tasks with which digital computers have the most enormous difficulty. To get a computer to do anything even remotely interesting requires the painstakingly precise combination of a very great many logical instructions. When it comes to really interesting tasks like pattern recognition, the complexity and ingenuity of the logic required becomes horrendous. For example, to write a computer program that could identify a human face, whatever the angle, whatever the lighting, whatever the person's expression, whatever the person's age, is a software task that lies way beyond our present capabilities—yet it is, of course, one of our own most basic and most taken-for-granted human skills. It is very difficult to isolate the principles which would be needed to form the basis for such a program because, when we think about how we recognize a face, there is actually nothing to think about. For us, the process is entirely intuitive. We either know a face or we do not. The logical processing which presumably underlies our facial recognition skills is completely hidden from us. Of course, we have other basic, similarly taken-for-granted skills which are just as remarkable: our capacity to use language, our capacity to see in three dimensions, our capacity to hear music—that is, our capacity to hear music as music rather than just a collection of sounds. Although we learn these skills to a considerable extent through experience, in each case there is an inescapable requirement for some kind of fiendishly ingenious bootstrap software to come pre-programmed into our brain, and therefore—if science is to be believed—further and more deeply pre-programmed into our DNA, in that simple four-base code.

To my mind—which is, for a good part, the mind of a programmer since software engineering is my original profession—this is an utterly astonishing thought. And it is made even more astonishing by the fact that these programs are so effectively able to combine robustness and reliability with plasticity and flexibility, characteristics which do not sit at all well together in the digital world of silicon-based computers. The point is that these qualities of our own carbon-based analogue computer can be no lucky accident. They must form an integral part of the overall program design, a vital part, and almost certainly a bulky part. In terms of a program's logical load, the price of features like robustness and plasticity is considerable. These are extremely expensive luxuries. The brain's ability to retain its integrity in spite of serious, sometimes quite appalling, damage is a phenomenon that has no parallel in digital computing. To think of sticking a crow-bar through a computer and still finding it able to perform in a useful way is ridiculous, yet the human brain has amazingly survived just such damage. This goes against the grain of everything we understand about logical systems. Just one loose hardware connection, or one wrong software instruction, is all that is needed to crash a digital program—just one tiny misplaced bit of information, a single 'one' misread as a 'zero'. To protect a system from that very simple kind of mishap requires considerable ingenuity of design. The brain's ability to reallocate its resources and recover from most, although not all, local failures is an astounding feat of systems programming. It compounds by a further order of magnitude my astonishment at the thought that the seed of all our brain's three-dimensional programming is originally programmed into just a few molecules of DNA. If it were not for the fact that there is simply no alternative rational explanation, few scientists, I think, would be prepared to believe it possible.

We have looked in detail at the biological role of DNA for the very good reason that it represents the logical point around which the whole scientific theory of origins revolves. The theory of evolution, in its simplest statement, suggests that simple forms evolve into more complex forms as a result of chance errors occurring in the DNA replication process. One of the millions of bases will be copied incorrectly to effect a minute change in the genetic instructions or the protein database. Crudely speaking, the theory goes that this tiny alteration to the genetic blueprint will occasionally prove beneficial rather than deleterious, in which case the offspring will be better adapted to their environment, and therefore more likely to survive, and survive longer, and are therefore more likely to produce offspring themselves, a proportion of which will inherit the mutant string of DNA, to be further passed on to their offspring and so on, until, in time, the mutation spreads throughout the whole population. The theory has a tremendously powerful simplicity, indeed a tremendously enticing simplicity. It almost *demands* to be believed.

At first glance, perhaps the unlikeliest feature of the scientific theory is that the DNA copying mechanism is designed to *prevent* the very errors which are supposedly the sole source of all nature's abundant creativity. However, this strange and seemingly counter-productive arrangement cannot really be avoided. For reasons that should be obvious, it is vital that DNA be able to replicate itself extremely accurately. Particularly during the billions of cell divisions of morphogenesis, it is absolutely essential that there is no violation of the DNA's program. There are, in fact, molecular mechanisms which are constantly maintaining the integrity of a cell's DNA, data security molecules which keep the twisted chains zipped tightly together to protect the genetic information from corruption. Copying errors have to be, and indeed are, incredibly rare, and will usually be limited to a change in, or a deletion of, a single base, or a very simple linear transposition of a few bases. The only time DNA undergoes any really major change is during sexual reproduction, when the DNA of the mother germ cell is scrambled with the DNA of the father germ cell. But this is only a swapping of genes, a shuffling of the functional sub-routines into which each DNA string is divided. This is reorganization rather than origination. Although it can allow the activation of mutant genes whose expression was previously repressed, or creatively bring together a number of mutant genes which only find useful expression in combination, the fact remains that however fortuitously the genes might get mixed in the fertilized germ cell, any truly creative morphological change, for better or for worse, is ultimately to be traced back to a small number of random and very slight DNA copying mistakes.

This brings us to the most commonly articulated problem of the standard theory—the question of evolutionary novelties, those incredible organs and mechanisms with which the natural world abounds, that either work as a perfectly integrated system, or do not work at all. The problem is that, in the case of an adaptation whose function is dependent on the perfect co-operation of a large collection of highly specialized components, it is extremely difficult to envisage the nature of the vast number of intermediate steps through which it must have passed in its evolutionary development, *each* of which has to be capable of a perfectly integrated function, and *each* of which has to represent a significant, selectable improvement over its predecessor. Wondrous examples abound throughout the natural world, but we will stay close to home and point to the classic example offered by our own eyes.

We have already talked a little about the amazing software processing that creates our visual imagery (see Chapter 3). There is little that can be said about the hardware which cannot be appreciated by simply looking in a mirror. In the masterful engineering of the eye we see revealed nature's most dazzling gem of creative virtuosity. The obvious wonder is the way in which the components of the eye—retina, lens, iris, cornea, with the support of a host of other precisely adapted mechanisms—all work together in perfect synchrony as one integrated whole. And the unavoidable implication is that they can only have evolved together in perfect synchrony, as one integrated whole. But a far greater wonder still is the way in which this immaculate hardware works together with the software. And again the unavoidable implication is that they too can only have evolved together in perfect synchrony, as one much more complex and even more closely integrated whole.

It is this introduction of the software side of the equation which raises the most profoundly troublesome questions. The chances of a hugely complex non-linear program being improved in its function by a simple random corruption of its source code are, to my mind, ridiculously astronomic. The chances of such a modification being accompanied by an appropriate hardware modification are surely so cosmologically remote as to be entirely discounted. Yet we are asked to believe that this is actually a very common happening, with many thousands of such tandem advances having taken place to create our visual system, as well as, quite independently, other technically very different systems across many other species.

While on the subject of evolutionary novelties, I must mention the most perplexing creation of all, and the very first one that we know: the DNA replication system itself. It has often been said, rightly I believe, that the gap between a soup of amino acids and the simplest unicellular organism is far greater than the gap between that same unicellular organism and ourselves. DNA is the software which organizes the running of the cell and the building of new cells, but it is quite useless without the hardware of the cell already being there in the first place. DNA replication and transcription relies on all sorts of complicated molecular mechanisms, mostly based on proteins which are unstable outside the precisely controlled environment of the cell. All those mechanisms have to be exactly locked into place in order for the DNA software to run. It is all or nothing. With a vivid imagination, science has proposed an evolutionary scenario to take us from the single cell to the human being, but it can offer no such scenario to take us from a soup of organic chemicals to the very first autonomous self-replicating system.

For me, there is no acceptable rational answer to the immense problems posed by evolutionary novelties. Take a look at a 'simple' spider's web or a bird's nest and think about how you might, as a spider or a bird, set about their construction. The more I think about the problems posed by just the very first steps, the more and more wondrous these structures appear to me. They point to the use of an astonishing range of creative, problem-solving skills. Yet, these incredible instinctive abilities are supposedly hard-wired into just a few molecules of DNA, mechanically unfolding—with seemingly unfailing precision—during the morphogenetic process. I just cannot stretch my mind to see how such magnificently ingenious feats of hardware and software real-time systems engineering can have come into being through a series of simple and entirely accidental mistakes in the DNA copying process. And I can only attribute the fact that so many scientists can to some kind of numbness to numbers. They can somehow kid themselves into believing in the plausibility of the standard theory because they have no handle by which to grasp the mind-blasting complexity of the processes and the programming involved. Here is a wonderful analogy from Murray Eden, a professor of engineering at the Massachusetts Institute of Technology, which is far closer to the truth than any rationalist would dare to admit:

> The chance of emergence of man is like the probability of typing at random a meaningful library of one thousand volumes using the following procedure: Begin with a meaningful phrase, retype it with a few mistakes, make it longer by adding letters; then examine the result to see if the new phrase is meaningful. Repeat this process until the library is complete.[5]

This analogy can be made even more strikingly accurate by emphasizing that not only does each phrase have to be meaningful in itself, but meaningful also in the context of the paragraph, which itself must be meaningful in the context of the chapter, which itself must be meaningful in the context of the book as a whole. The more complex the work, the more difficult it becomes to generate meaningful phrases. In the case of a very precisely worked treatise, each new phrase will be almost exactly defined by the meaning of what has gone before. Out of countless billions of possible phrases, only perhaps a few thousand will make any sense in this very detailed context, and most of those will be neutral in that they say nothing new. Only a handful of the sensible possibilities will actually further the work's line of reasoning. To sharpen the analogy yet further, we could replace the thousand volumes by a thousand large computer programs. Instead of adding new words, we now add new logical instructions. Beginning with the simplest possible working program, we reload it with a few mistakes, add a few extra random instructions, and run the program to see if it will work, and also to see if it will interface properly with all the other programs that are evolving concurrently. We repeat this procedure until our full complement of programs is up and running together. A programmer will instantly recognize this procedure as even more unlikely and cumbersome than the previous one. Again, the more complex the program, the more difficult it becomes to generate meaningful sets of instructions. In the case of a very precisely worked program, each new logical instruction will be almost exactly defined by the meaning of what has gone before. Out of the countless billions of possible program modifications, only perhaps a few thousand will make any logical sense, and most of those will be neutral in that they add nothing new to the program. Only a handful, if that, of the sensible possibilities will actually further the program's functioning.

The parallel with DNA's complement of genetic programs should be obvious. Just like a computer program, each instruction that is encoded into the base sequence of the DNA chain must be precisely defined by its overall logical context. The astonishing coherence of every individual part of our whole organic system must necessarily be contingent upon an even more astonishing coherence of the individual instructions in the DNA's program. Every element in this program is an interrelated part of one superbly integrated functional assembly, combined together in *just* such a way that the entire make-up of our physiological being, both structural and functional, hardware and software, is supposedly packed implicitly within the DNA's base sequence. Against the backdrop of this incredible level of logical synergy and inter-dependence, implicating tortuously convoluted loops of internested molecular feedback and feedforward, the proposition that a procedure of the kind described above has conjured up millions of individually astonishing evolutionary novelties in so short a time is untenable. The scientific picture of evolution is surely incomplete. The elements that we have discussed cannot be made to fit together in quite the tidy, rational way that science would like.

The discovery of DNA and the cracking of the genetic code has actually made the evolutionary process more rather than less mysterious. To be able to appreciate all the amazing details of the scientific narrative, to feel the power of that narrative, and, in that thorough intellectual understanding, still to be compelled to reject it as incomplete—that is to be invaded by a rather special sense of wonder and awe. And, indeed, it is some feeling for this intellectually sharpened spiritual sensitivity that I have wanted to capture in this chapter. The inadequacy of the scientific story cannot be pinned down to any one particular point. The problem lies with its entire framework. Its rationality. Under the rationality of reason, being is seen as a product of evolution rather than as the essential *context* of evolution. The standard scientific theory runs into brick walls because it is trying to see the subjective pole of being as having evolved out of the objective pole, rather than seeing both poles as having evolved with each other, and as having involved *into* each other. We have to return again to the idea that from the very beginning of evolution there has existed some kind of subjective rationality of wholeness offering context and meaning to the objective rationality of the parts.

According to science, though, the ultimate origin of our existence is to be traced back entirely to the Big Bang—the explosive birth of all that is. By mathematically conjuring the universe out of literally nothing, the Big Bang theory of creation has acquired an almost sacred status for modern scientists. It goes almost entirely unchallenged. But its underpinnings are far less sound than we are generally led to believe. There is a gaping crack in its foundations which is almost invariably ignored. The flaw has all to do with the fact that there is no absolute physical scale in the universe. In measuring any quantity, be it space or time, matter or energy, all we can do is make comparisons against some previously prescribed reference quantity. The theoretical foundation for the Big Bang model comes from the general theory of relativity, the mathematics of which suggests that spacetime has to be expanding. The trouble is that when we have no absolute basis of scale, there is no way of understanding what the expansion of spacetime really means. The popular picture of the exploding universe is of an inflating balloon on which are painted a large number of stars and galaxies. As the balloon inflates, so all the stars and galaxies are seen to move away from each other. And this concurs with experimental evidence. But, as the universe expands, the scale that is used to measure distance would presumably expand also, thereby denying us any perceptual awareness of the expansion. From the point of view of an observer on the surface of the balloon, there is no expansion. Subjectively, the concept is meaningless.

Modern cosmologists extrapolate backwards from the expansion of the universe to locate an 'event' from which space and time themselves have sprung forth into existence. This is the Big Bang. But, in the absence of any absolute scale, the possibility surely exists that we can extrapolate backwards indefinitely. Mathematically, we can spatially reduce a pattern of information any number of times without fear of losing its meaning. Even after an infinite number of reductions, the information content—the pattern—is still perfectly preserved. I am reminded here of the wondrous computer-generated representations of the Mandelbrot Set. This mathematical object is exquisitely simple in definition, but infinitely complex in structure. It can be logically defined with a few symbols, yet it contains its own infinitely rich universe of fractal pattern. At every level of resolution, new and unique detail is revealed within the same scale and general structure of patterning. The Mandelbrot universe is self-similar at every scale.

The concept of an explosive beginning to our universe is almost entirely dependent on the very speculative assumption that the measuring scales which determine all the fundamental atomic quantities—the 'constants' in the descriptive equations of physics—have remained totally unchanged throughout the evolutionary history of the universe. Modern cosmology requires that the measuring scales which we use today are valid in an *absolute* sense. It requires that they be unaffected by the evolution of the universe in which they are embedded, that they exist completely independently of its large-scale structure, indeed predetermined before the Big Bang even brought the universe into existence. It is an assumption which is very much in opposition to the spirit of relativity.

The main empirical evidence for the expansion of the universe comes from Edwin Hubble's famous discovery of the relationship between the distance of a stellar object and its red shift—the degree to which its light is shifted toward the red end of the spectrum, and taken to indicate that the object is receding from us. Hubble discovered that the more distant the stellar object, the greater its red shift, and therefore the faster its velocity of recession. The interpretation is that the universe is expanding. Every object in space seems to be moving away from every other. Reverse all those motions and we find every object traceable to a common origin—the Big Bang. But Hubble's relationship can be interpreted in terms of my alternative vision of the expansion of the universe. It could be revealing the fact that it is our measuring scale which is expanding. Although our expanding universe never appears, from the inside, to be any different in general appearance, the light of a distant star or galaxy represents a fossil record of an earlier time, perhaps when the universe was relatively smaller, or, more properly, had a relatively smaller scale. It could be that the expansion of the universe has effectively stretched, or red-shifted, the ancient light of distant stellar objects. Here is an example of a universe which is self-similar at every epoch.

The mathematics of general relativity suggests a model which is finitely bounded in space and time and contains a point of creation. However, from a subjective perspective, from inside the universe—which is the only perspective to have any real meaning—there need not necessarily be such a point of creation. Without an absolute scale, it becomes possible to extrapolate backward in time indefinitely. The universe then has an infinite history. It has neither a beginning nor an end. Although from the 'outside' this infinite universe is seen to be finite in time, with a perfectly well-defined birth and death, this perspective is utterly fictitious. It is a mathematical conceptualization, an abstraction. It is without reality. There is no such objective, Godly perspective. Remember, the universe is a Mind. There is *no* outside. There is only the internal, subjective perspective offered from the eternal 'now'.

The real point of this discussion is not so much to answer the question of which model of the universe is the more valid, but to stress that cosmology cannot evade this kind of ambiguity of interpretation—and that in no other branch of science is there more creative scope for shaping observational results to fit one's preconceptions. Cosmologists wish to capture the universe objectively, as an entirely physical structure, and therefore choose to picture it geometrically from an external perspective, like God looking on in admiration of His handiwork. But the truth is that this perspective is a fiction. It just is not possible to step out of existence to see what Being looks like from without. There is only the within. Even if the physical universe did have a Big Bang origin in finite time, the *potentiality* for space and time must still have been present in some prior sense. There must have been some prior rationality of being. But there is a certain ridiculousness to this kind of discussion. To inquire into the creation of the universe is to engage with a concept that lies way beyond the conceptualizing capability of the human mind, a concept that cannot be even remotely touched by reason, a concept that cannot even be touched by religion. Even God could not hope to embrace the rationality by which existence has existence, the ultimate mystery through which his own divine being had come into being.

Here, in its hopeless inability to embrace the humbling mysteries of time, as well as in its fumbling grasp of the experiential interaction of mind and world, we can see most clearly exposed the inadequacy of the rationality of traditional science. We surely need to look beyond the dogma which insists that all phenomena are ultimately to be reduced to the interactions of elementary particles, governed by just a few timelessly valid physical laws. There seemed promise once that the universe would indeed yield to such a simple reductive analysis, but although the anonymous institution of science is stubbornly disinclined to let go of the hope of winning such a glittering prize, it is dawning upon an increasing number of individual scientists that this promise has now evaporated. Their insight has grown deeper. There is more to the universe than can be explained away purely in mathematical terms. The traditional scientific story can no longer be considered reasonable as a total description of the universe. By trying to stretch its rationality too far, science has actually become irrational. The rationality of reason just is not powerful enough to embrace the ultimate mystery of creation. We need to move on from our rational story to a suprarational story.

But, again, any such suggestion that there exists some higher-order suprarationality remains anathema to the institution of science, which clings pertinaciously to the belief that an expensive enough programme of atomic vivisection will eventually reveal a mathematically encapsulable 'Theory of Everything'. The immensely powerful and persuasive objection to evolution as involution is that it complicates and obscures what to science still promises to be a simple and logical story. We trade a describable theory for an indescribable one. The big problem with our suprarational context, and why it is so intellectually unattractive, is that it just cannot be modelled. Even in the age of the new physics, we still hanker after solidity. We like to picture things. Most of all, we like easy answers. For the sake of elegance, not to mention intellectual pride, science would love its rational story to be the whole story. And there are a great many who firmly believe that it is indeed the whole story, to the extent that there has grown up a colossal prejudice, a prejudice which is presently vitiating scientific objectivity—a monumental mass blindness to the true depth of the perplexities that we have just explored.

The trouble is that scientists make such an enormous emotional and vocational investment in the rationality of reason. They bank away so much of their time and energy. They *have* to numb themselves to the unconscionable improbabilities that are implicit in their rational story of evolution, because their very respectability and livelihood as scientists depends on it. I can understand this because there are times when I become afflicted with this same number numbness. As an annoying remnant of my cultural conditioning, I find that the desire to see the whole as a simple collection of rational parts is deeply insistent, sometimes irresistibly insistent. There are times when my consciousness gets stuck in the rational world of the left hemisphere, the persuasive power of the rationality of reason consuming my spiritual sensitivity. In moments of epistemological weakness, it is easy to be moved by the siren call of the simple picture that is offered by a stand-alone mathematical rationality. This is one of the reasons why I felt it was necessary to spell out my objections to the scientific story in so much analytical detail. But analysis has its limits. Although the prejudice harboured by the rational mind has to be challenged through analysis, ultimately, only an intuitive feeling can encompass the need to sacrifice the inviolacy of scientific rationality. With the analysis of this chapter, I hope I have helped to bring this feeling a little more clearly into focus.

Since the context of our suprarational theory of evolution is being, transcending the categories of object and subject, there are inevitably very great difficulties of presentation to be overcome. As I have just implied, we are faced with that same old problem, the ineluctable hold of the idea that we inhabit a world of objective solidity. Our vision cannot find escape from the concepts of spatiality and substantiality. Our thought is constrained by an infuriating inability to see an object as anything other than an independent reality which consists of lots of other very much smaller, similarly independent realities. The view that these elementary realities are the only true ones is etched very deeply into the modern consciousness. This description obeys what we shall now call the rationality of the Many. This is the familiar rationality of science. We have seen, though, albeit hazily, that these independent elementary realities are not really so independent. Quantum theory has shown us that their reality is somehow coupled to the act of observation. Although there is indeed a sense in which they exist as separate realities, there is another, complementary sense in which they are inseparable from the wholeness of reality. Another rationality is needed. We shall call this second rationality the rationality of the One. Obeying an alogical calculus, the rationality of the One is tantalizingly elusive to analytical description. Consequently, the language of our suprarational story will necessarily be as much poetic as scientific.

More than just representing complementary, mutually exclusive modes of description, the rationalities of the One and the Many actually define complementary, mutually exclusive modes of being. In its Oneness, the being of the Mind that is the universe is present to the being of every aspect of its Manyness. However, this is not to suggest that every part has some kind of mind. It is to suggest that every part is a participating component in the Mind of the Universe. Every part is informed by the whole, in a way which we have yet to even begin to understand. The rationality of the One co-exists with the rationality of the Many. The higher-level principles which are embraced by the rationality of the One organize the parts within the constraints imposed by the rationality of the Many. The rationality of the One functions in such a way that there is no contravention of the lower-level physical laws of nature. We could say that the rationality of the Many operates at the hardware level of Mind, while the rationality of the One operates at the software level. However, these hardware and software levels are tangled in a way which totally defies comparison with anything from the world of computing. The whole is composed of parts, but those parts are composed by the whole— an unbroken, unbreakable epistemic circuit, the original paradox of existence.

Within this mystical context, evolution is still to be seen as a stochastic, heuristic process, except that the rules of engagement are widened. Instead of creativity being traced back to a random shuffling of the objective parts, we now see it as the product of an inherent playfulness between the whole and the parts, between the One and the Many. This brings us back to the eternal *Tao*. We can identify the playful tension between yin and yang—what the ancient Chinese understood to be the primal rhythm of the universe—with the playful, experimentative tension between the One and the Many. It is in this dialectic that we find the spirit of evolution. As the parts become increasingly integrated in structure, so the whole is able to inform upon the parts in an increasingly profound way, so the whole becomes more involved in the organization of those parts. Evolution is a bootstrap process, and an accelerating one, by which the One and the Many pull each other up into an increasingly coherent existence. Evolution is seen in terms of a growing involvement of the One in the Many, and the Many in the One. This is involution.

In a way that is frustratingly difficult to articulate, I find a deep feeling of rightness about this nebulous image. We normally find nature to be so sublimely efficient that it would seem tantamount to a careless flaw in an otherwise perfectly designed universe if there were not some kind of feedback mechanism by which the whole could influence the evolution of the parts. As we found with quantum theory, what makes it so difficult for us to model the mechanics of this kind of epistemic feedback is, once more, that haunting context of wholeness, transcending the object/subject categories of our cognitive processes. Quite inescapably, we are part of that wholeness. There is simply no way in which we can break out of the circle to see the whole picture. The observer and the observed are joined by the rationality of the One, a connection fated never to fall within the embrace of reason.

Ultimately, though, the fine details of our suprarational story are of far less importance than its overall texture. This story is not about explanation. It is about the *mystery* of evolution. And it is this aspect which we should be emphasizing in our children's education. Indeed, I believe that the presentation of evolution as involution should form the nucleus around which the major part of the school curriculum is organized. In this way, the teaching of science and the humanities would be charged with a new vitality and relevance. The presentation of evolution as involution would provide a means of complementing fact with feeling, breathing value into the physical and biological sciences as they serve to tell the story of our creation. The principal aim would be to inspire students with the mystery of evolution. Students would be encouraged to enter into the mystique of their evolutionary origin, to find in that ineffable beginning an awareness of their own numinous depths and possibilities. Such an education would therefore be religious as well as scientific. To quote the words of Albert Einstein:

> The most beautiful emotion we can experience is the mystical. It is the sower of all true art and science. He to whom this emotion is a stranger ... is as good as dead. To know that what is impenetrable to us really exists, manifesting itself as the highest wisdom and the most radiant beauty, which our dull faculties can comprehend only in their most primitive forms—this knowledge, this feeling, is at the center to true religiousness. In this sense, and in this sense only I belong to the ranks of devoutly religious men.[7]

The invocation of the rationality of the One introduces a spiritual dimension to our story of our origins. Evolution as involution communicates a sense of the immanence of the divine throughout the universe, the numinous presence of the One to every articulation of the Many. And this represents the most immediate significance of our suprarational story. For the authentic experience of mystery is an archetypal human experience, the inspirational impetus to our religious instinct. To have this experience denied by a pseudo–explanatory narrative of creation is to be denied a vital aspect of our humanity. Only by recapturing a common sense of the mystique of creation can we hope to heal the autistic spirituality of the Western consciousness. To be aware of our own mystical presence in the world is to be able to identify, through the rationality of the One, with the evolutionary process itself. It is to find a sense of belonging to the Earth as the mother of evolution. To recognize in ourselves the very same divine mystique that is present in the being of the world is to enter into a passionate global communion. To identify with the evolutionary process is to find our soul in the sacred soil of the Earth. The land we live upon becomes hallowed, to be venerated, respected, loved, celebrated. To experience the meaning of evolution as involution is to feel the creative rhythm of the universe resonating in the radiant currents of our own being. To open ourselves to the narrative of this story is to feel our own votive involvement in the narrative. It is to enter into the narrative, a sub-text, influencing and shaping the narrative flow. It is to understand that each one of us is a poet, empowered to write our way into the Earth's evolutionary story. This story is *our* story, our integrating story, inspiring commitment and community and communion, the communion of involvement.

10

The Mythology of Survival

O UR DEFILEMENT of the Earth is a direct consequence of the defilement of our human spirit. Depriving ourselves of a soul has meant depriving ourselves of our human dignity. And this is what the adoption of the rational mythology of traditional science has surely done. We have lost our dignity to reason. Taught only to see with the eyes in our head, we have forgotten how to see with the eyes in our heart. This is our spiritual myopia. The eyes in our head are not equipped to see the profound organicity of our relationship to the Earth. Through the eyes in our head, we are only informed by the rationality of the Many, revealing the individuality of our human presence in the world, the singularity of every human articulation of being as a unique locus of involution. Only through the eyes in our heart can we be informed by the rationality of the One, revealing the *universality* of our human presence in the world, the identity of every human articulation of being with every other, and with the being of the Earth, and ultimately with the mystical being of the universe. As witnessed so terrifyingly in the political affairs of the major nations of the world, a rational belief system fails hopelessly to encourage rational decision-making. Ultimately, the rationality of the Many can only foster irrationality, because its solitary viewpoint is that of the part. The reality of the whole is secondary to the reality of the parts. This is the heartlessness, the spiritlessness, of the Western perception, our vanity. The rights of the Earth are seen to be secondary to the rights of an élite and spoilt minority of its present-day human inhabitants.

A spiritually healthy life is a life which affirms the individual *and* the universal aspects of our being. Authenticity is to be found in a dialectic between the individual and the universal, in a striving for a balance between our own personal interest and that of the whole. This is another way of describing the goal of Being, the ultimate rainbow of existence, destined forever to be pursued by the human spirit and, of course, destined forever to elude it. For Being contains an inherent opposition, a contradiction which is intrinsic to the essential nature of our existence. The values of the individual and the universal express an inescapable conflict of interest. The good of the individual is continually at odds with the good of the whole community. But mostly we cannot afford to be consciously aware of such conflicts, because our life would then become consumed by them. Internally, we set some kind of threshold which censors our awareness of the more uncomfortable contradictions in life.

For example, most people who go walking in the hills will take their litter back out with them, but it is only a minority who feel compelled actually to pick up the litter that has been left so carelessly by others. Picking up other people's litter is too much effort, and too annoying—because it is painfully reminiscent of the values that we have come into the hills to escape. Sometimes I pick up litter in the hills; at other times, though, I have to admit that I run straight by. I censor my own awareness in order that I can enjoy my walking or running without interruption. Once you pick up one piece of litter it is very difficult to stop looking for more. Instead of enjoying the scenery, you find your eyes being drawn, resentfully, toward the ground, searching for the detritus of our throwaway culture.

Our contemporary crisis of value is really all about where the threshold of our awareness is set. That in some cases it is set so low that people feel free to litter the hills with their rubbish is desperately sad, both for them in their pathetic self-centredness, and for the world at large. But the crisis doesn't really lie with this visionless minority. It properly lies with those of us, the majority I feel, who are informed by the rationality of the One to the extent that we would not drop litter, but not always to the extent that we feel compelled to pick up the litter of others. The litter is, of course, a metaphor. Looking after our own litter symbolizes the acceptance of personal responsibility. Picking up other people's litter symbolizes the acceptance of global responsibility. Taking responsibility for our own individual actions is vitally important but on its own, sadly, not enough. We cannot avoid also being responsible for the actions of others. In order to embrace an authentic mythology of progress, we first have to summon up the courage to face our global sense of responsibility.

The conflict between the rationalities of the One and the Many has come to be known as the tragedy of the commons, from the name of an article written by Garrett Hardin, an evolutionary biologist. The 'commons' refers generally to any shared resource. Hardin's classic example is of a shared enclosure of grazing land, traditionally called a common, upon which each herdsman in the community has the right to graze his animals. The problem is that if each herdsman were to allow all of his animals to feed upon the common, the land would quickly become overgrazed and useless. Yet because it is free, each herdsman is tempted to graze as many of his animals on the common as he can, certainly at least as many as his neighbours. Each herdsman's desire to make sure of his fair share eventually guarantees that no one gets any share at all. This is the tragedy of the commons. Replace herdsman with nation and the common with the Earth, and we have the tragedy of our beleaguered planet.

In his wonderfully wide-ranging and thought-provoking *Metamagical Themas: Questing for the Essence of Mind and Pattern*, Douglas R. Hofstadter discusses the tragedy of the commons with his usual polymathic wit, coming up with some fascinating perspectives. Most importantly, he argues that in a 'commons' situation, the rationality of the individual person, if it pays no heed to the global rationality, is actually an irrationality. In Hardin's example, the irrationality is obvious. If the herdsmen were simply to exercise a little short-term restraint, they could guarantee the long-term health of their common pasture. As it is, their individual greed turns and acts not just against the global good, but against the individual good too. The irrationality of our own situation on the Earth is no less obvious, except that the scale is so great and the implications are so awful that we cannot bring ourselves to acknowledge it—nor, of course, do anything about it.

We each decide that the tiny little bit of restraint that we could exercise individually would be of such little consequence on the global scale that there is no point bothering. It wouldn't make any difference. And, anyway, why should we bother when none of our neighbours are showing any restraint? Why should we miss out on the bonanza of consumerism? As Douglas Hofstadter so neatly puts it, 'Apathy at the individual level translates into insanity at the mass level.'[1]

As a vehicle for exploring the relationship between individual and global rationalities, Hofstadter chooses a variant of a simple game called the prisoner's dilemma. It is played for points between two people who have no communication with each other, and no knowledge of each other. Each player is given the choice of co-operating or defecting. No more than that. Co-operate or defect. Now, if both players choose to defect, their pay-off will be one point each. If one player chooses to defect and the other to co-operate, then the defector's pay-off will be five points and the co-operator will get none. On the other hand, if both players choose to co-operate, their pay-off will be three points each. The aim of the game is to acquire as many points as possible. Now, imagine that you are one of twenty-one people to have been invited to take part in a round robin tournament of this prisoner's dilemma game. In fact, Hofstadter did organize such a tournament, with twenty players. You are the twenty-first! You are asked simply to make a choice between co-operation and defection. That one choice will then be matched with the choice of the other twenty players. At the end of the tournament, you will receive the sum total of points due from the twenty pay-offs, as calculated in the manner described above. You have no idea who the other players are, and have to assume that you never will. And remember, the idea is to come out of the tournament with as many points as possible. Pretend that you are playing for money—£100 a point!

Well then, what is your choice? Are you to co-operate, or are you to defect? It's a tough decision. The situation is full of the most horrible paradoxes. There is no rational strategy. The dilemma is that collectively it is in the best interests of everyone to co-operate, while individually it is in the best interests of everyone to defect.

Let us attach some numbers to this to make the point as clearly as possible. If each player were to co-operate, everyone would receive £6,000, twenty lots of £300. However, the choice to co-operate has to be made in the knowledge that defection is *always* more profitable. If everyone else has co-operated, you can clear £10,000 by defecting yourself. On the other hand, if each player were to defect, everyone would receive just £2,000. And if you were unfortunate enough to be the sole co-operator among twenty defectors, you would end up with precisely nothing. Put this book away for a while and think very hard about exactly what choice you would make.

I hope you have put some hard thought behind your decision. Did you co-operate? If you did, I have to suggest that you made a very strange choice, for by co-operating you were guaranteed to win fewer points than by defecting. Suppose you were told that you were the last person to make a choice. Would you still co-operate? Since everyone else has already posted their decision, it would make no sense at all to do anything but defect. Yet, why should knowing that your choice is the last make any effective difference? You have no contact with the other players. You might as well assume that everyone else has already decided anyway. There is an overwhelming case to be made for defection. Do you want to change your mind? Or perhaps you chose to defect in the first place. If you did, I again have to suggest that you made a very strange choice. Since there is such a strong case for defection, it is likely that everyone chose to do exactly that, receiving £2,000 each. However, looking from the outside, from a global perspective, at the twenty-one of you collectively, your unanimous choice to defect appears as an act of unalloyed folly. You have each short-changed yourself to the tune of £4,000—the extra cash you would have won had you each co-operated. There is an overwhelming case to be made for co-operation. Again, do you want to change your mind? What is your final choice?

So, let us see how you fared in Hofstadter's tournament. He found fourteen defectors and six co-operators, with some fascinating explanations to justify people's choices. Adding your choice in to his matrix of games, the result is that if you chose to co-operate you pick up £1,800, while if you chose to defect you pick up £4,400. I will leave you to draw your own conclusions.

Let us bring the discussion back to everyday life. Imagine that we suddenly experience a phenomenally cold spell of weather, of Arctic bitterness. The demand for electricity would rise dramatically, everyone cranking their heating up to the maximum in an effort to keep out the cold. Add to that lots of extra hot meals and drinks, and the demand for power soon overtakes the electricity company's capacity to supply. An announcement is made which requests people to turn their heating down, otherwise there will have to be indiscriminate power cuts. Either everyone reduces their demand by a little, or some people will have to go without power completely. And that could include you. Do you reduce your consumption? As with the tournament above, it is easy to believe that there is nothing to gain by co-operating. Your little bit of electricity is not going to make any difference, so, if most people have co-operated, you might just as well keep your heating right up and stay snugly warm. On the other hand, if the majority have defected and kept their heating up—selfish, thoughtless people—there is already a chance that your power will be cut off, so you might just as well keep yours up too and make the most of it while it lasts. No one is ever going to know. Of course, if *you* think this way, it is a very good bet that a lot of other people will be thinking the same way. And that is why you suddenly get blacked out—and frozen out. Individual apathy snowballs into global insanity.

The decision to keep your heating turned up full, like the decision to defect in our tournament, is perfectly rational. Yet, as we have seen, it is also irrational. There is a need to refer to some kind of higher rationality. The suprarationality of the One is such a rationality, and ultimately this is the rationality to which we must look eventually to solve our ecological problems. The trouble is that it is inherently subjective. It represents an intuitive value, a *personal* value, communicable only through experience. An appeal to the rationality of the One is not going to get the immediate response that the urgency of our predicament demands. In the short term, we need to appeal to what Douglas Hofstadter calls super-rationality. The decision to turn your heating down a notch, putting on an extra pullover instead, is a superrational decision. Hearing a rumour that there is going to be a shortage of some commodity, coffee for example, and therefore buying a little bit *less* than normal, rather than stocking up and helping to bring about the rumoured shortage, is a superrational decision—because, by taking account of the collective good, your superrational choice will eventually be reflected back to work for your own good. You hope. The fear is, of course, that those people who are addicted to their caffeine in a serious way will panic and hoard supplies, clearing the shelves so that next time around you will go short. And it is that fear which exerts such a strong grip over our mind, making us want to buy in bulk too, in order to guarantee our own supply. Just as in our prisoner's dilemma tournament, the overriding thought is that you can only be worse off as a result of co-operating. As we saw previously, the rational case for defection *seems* to be overwhelming. But defecting is not rational. It is thoroughly *irrational*.

Our difficulty in deciding between co-operation and defection is the result of our incomplete knowledge. Because we have no means of communicating with our fellow players, because we have no means of knowing who they are and what they might be thinking, we don't feel able to trust them to co-operate. So we defect. And so do they. And the result is that we are all worse off. Yet we don't actually need to be able to talk to each other to communicate. We can communicate through a common principle, a common value. The greater part of this book has been given over to the idea that we are subjectively connected through the suprarationality of the One. This is the interior aspect of our green rainbow, the value of Being, our existential bridge between the personal and the universal. But we are also connected objectively through the value of superrationality. This is the exterior aspect of our green rainbow, the objective representation of our more profound, subjective connectedness, as revealed by our decision-making and socio-political involvement. If we were to live in a culture which honoured and respected the value of superrationality, a culture where sanity prevailed over insanity, we would be much more inclined to trust our fellow players. It is this value that we have to promote by example. Trust engenders trust. Sanity promotes sanity. To play in a superrational way, regardless of the way in which our fellows are playing, is to assume a global responsibility for our actions. This is to play in an infinite rather than a finite way. Superrationality is the rationality of the infinite player of the game of life, the rationality of survival.

Most of us are superrational to a degree. To illustrate the point, let us return to our earlier example of litter. Say you were on a touring holiday in a foreign country that you would probably never visit again. You hike into the hills for a picnic. After eating and drinking too much, you fall asleep in the sun, and when you eventually awake, tired and lethargic, you don't much want to clear up the lunch mess and pack all your rubbish away. Why not just walk off? You will never be back to this spot. You don't know anyone here. No one knows you. Yet, although a few people do sadly just walk away from their rubbish, the majority of us do not. Our conscience speaks to us. Understanding that we would not want to chance upon such a mess ourselves, we do what we wish other people would do and leave our picnic spot just as we found it. We are joined with our fellow hikers by a common superrational value. However, if we were to come across the mess that someone else had left, to be rightly true to that superrational value we would be compelled to pick up the rubbish on their behalf. We have to take their responsibility upon ourselves. If we leave this mess, it is likely that other not so superrational hikers will also leave theirs, thinking that their little bit of mess won't make any difference to the mess that's already there. Our values are part of the value context of our community. Only by living the value of super-rationality can we expect our fellow players of the game of life to live it also. The more superrational we become, the more superrational we can expect our fellow players to be. We can only encourage people to infinite play with our own infinite play. We can only promote sanity with our own sane behaviour.

As I suggested a little earlier in this chapter, our crisis of value is all about where our threshold of superrational awareness is set.

Very few of us are superrational enough. We don't mind being superrational if it is not too inconvenient, but there is a threshold above which the commitment is too great for us to allow ourselves to be aware of it. We don't want our superrationality to interfere with our life in too big a way. But this simply is not good enough. We have to start taking global responsibility as well as our own personal responsibility. And we must accept that our life is going to change as a result—although by no means should it have to change for the worse. Far from it. To commit ourselves whole-heartedly to the value of superrationality is to begin to lead a very much richer life, a more authentic life. To quote Milan Kundera, from *The Unbearable Lightness of Being*: 'The heavier the burden, the closer our lives come to the earth, the more real and truthful they become.'[2]

The kind of commitment we are talking about here is somewhat deeper than that being shown today by the typical consumer with Green pretensions. As more and more people rush to climb aboard the Green bandwagon there has to be a strong suspicion that, for many, being *seen* to be Green is the most important thing. What looks like commitment is just a fashionable façade. To borrow the obnoxious terminology of the media people, Green is now sexy. The end of the eighties has seen the Green banner hijacked by the image-makers. Although the raising of the collective consciousness has to be regarded as healthy, the danger is that with so much emphasis being placed upon the virtues of Green consumerism, awareness is being diverted from the deeper and more urgent issue of values. Here is Jonathon Porritt, one of the very wisest voices of the Green movement, writing in *Resurgence*:

> The real issue, if people are seriously aspiring to cope with global warming, the desperate plight of the Third World, and the progressive deterioration of our global life-support systems, is not only how we might learn to consume better, but also how we might learn to consume *less*. And can this be done without sacrificing our quality of life?[3]

The answer is that we can indeed consume less without sacrificing the quality of our life, if the values that inform the quality of our life are dialectic rather than external, founded on who we *are* and what we *do* rather than what we *own*. Once we feel the emotional connection between ourselves and the Earth, we find that we do not want to spend our spare cash on a dishwasher. We understand that our quality of life is improved in a far more profound way if we invest our cash surplus in, say, a reforestation project. And then we can regard our chore at the sink each day as a meditation, a time of simplicity, a short interlude of serenity in our hectic life schedule, a few minutes to quieten the intellectual mind and reflect on our connectedness with the Earth's life-support systems, understanding that the trees we have helped to plant will be absorbing the carbon dioxide that was released into the atmosphere when the coal was burnt to generate the electricity that has heated up our washing-up water.

The power of the Green consumer to dictate to the industrial manufacturer is great indeed, and certainly an extremely valuable force in the market place. But our superrationality needs to go much further. The danger inherent in Green consumerism is that it will breed complacency and, worst of all, self-righteousness. People will think that their superrational duty can be carried out in the supermarket alone. It cannot. Most importantly of all, our superrationality has got to infiltrate the political arena. This is where global responsibility has most urgently to be assumed. Let us recall that original tragedy of the commons situation, concerning the common pasture. It is easy to see how the herdsmen could solve their problem. They live in the same community. They know each other. All they have to do is to draw up a convention which limits the grazing rights of each individual in order that, collectively, the grazing can be optimized. If necessary, the common could be policed and fines charged if any individual herdsman were to violate the convention. This deterrent would hopefully ensure that superrationality prevailed. The same solution is appropriate for us. Lacking a sufficiently widespread emotional understanding of our suprarational Oneness with the Earth, we have to enshrine our intellectual understanding of our superrational oneness in some kind of global convention—a covenant between humanity and the living Earth, a Bill of Rights for the Earth. Although there is much we do not yet understand, we have to make a start right now. We cannot afford the luxury of procrastination. Time is running out too fast. Each one of us has a superrational duty to apply all the political pressure we can to start making things happen.

We in the developed world cannot divorce our lives from those of the people in the non-developed world. We belong to the same human fellowship, living on the same land, playing in the same infinite game. That connectedness has to be given an economic expression. While we continue to indulge our consumptive binge, we cannot expect the poorer nations of the world to preserve their forests and limit their own industrial development. Their immediate and understandable priority is their people, not conservation. When your own children are starving, the ecological integrity of the Earth is of very remote and minor significance. And anyway, our lands were once covered by trees too. How is it that it is all right for us to strip our forest cover and reap the industrial benefits, but not for others? The remaining rainforests of the world are a natural resource far more valuable than oil or gold. They are the lungs of the Earth. They *have* to be preserved. And we in the West have a duty to pay the necessary economic price to guarantee their preservation. We must curb our profligate consumerism, not by denying the individual the right to consume, but by ensuring that the consumer pays in full the true environmental price of their consumption. Although it has to be admitted that it is not yet clear how such an ecological costing is to be calculated, nor how its payment could be globally distributed, what *is* clear is that this is the direction in which we have to go. These are the economic problems which we have to address, and our first objective in the political arena must be simply to establish their priority. We must begin by defining in detail the nature of the problems that we have to solve. The technical difficulties are, of course, very great—and beyond the scope of this book to explore in any detail—but they are not insurmountable. We have to find a way of expressing our global responsibility in an economic currency.

The capitalist world has proved beyond doubt the power and efficiency of a market-driven economy. The problem is that the market is informed by no value save that of the shareholder's dividend. The Earth simply cannot be trusted to the profit-tied hands of giant industrial corporations. That also has been proved beyond doubt. The power of the market must be put to the service of superrationality. Through a system of economic incentives and disincentives which symbolize the interconnectedness of people and planet, we will have to introduce far more central control to the market. Our superrationality will have to be encouraged by means of a programme of centrally planned fiscal persuasion, or, where appropriate, even a little fiscal bullying. And there can be no camouflaging the fact that it is going to hurt. Like spoilt children, we are going to resent having some of our material freedom curtailed in this way. But we had better start getting used to the idea. The reality is that authentic progress requires the sacrifice of some degree of individual freedom for the sake of the universal freedom in which the whole living Earth participates.

The saddest and most frustrating thing at the moment is that the longer we leave the present status quo intact, the greater the price that we will be forced to pay in the end. There is so much that we could be doing right now, obvious and simple things like recycling waste and ensuring that energy and natural resources are used more efficiently, but the market will not provide the initial capital. There is no ecological balance sheet that will reveal the underlying long-term cost benefits. It is madness to spend billions on massive new power stations while spending nothing on energy conservation. Even if we were to conserve just a tenth of the energy that we technically could conserve, we would still be able to close down power stations instead of having to build more. It is in this respect that we need to bring more central management to our political affairs. However, as certain powers are centralized in this way, so others will have to be decentralized. There are many areas where it is vital to give individual people more freedom of choice in their lives. The global strategy and policy-making that we need so urgently will have to be informed by local needs. It will have to work by providing a central legislative framework to which locally determined planning and decision-making must refer. Within that framework, the power of policy-making should reside as much as possible within each community, so that decisions are made by the people whom those decisions most directly effect. The details of how such a top-down/bottom-up power structure is best organized are not yet clear, but what *is* clear is that this is the direction in which we have to go. We need to give people the chance to exercise their global responsibility through more genuine freedom of choice—including, perhaps most importantly, more democracy, the democratic freedom of fair representation in a political arena purged of powerful vested interests.

Political parties need to be divorced from the institutional power-bases of society. Their funding needs to come from a central pool, indexed directly to their grass-root support among individual people—that is, the number of their individual supporters, rather than their collective, corporate wealth. In this way, new political movements, reflecting the changing values of individuals within society, have a route by which they can bridge the gap to the established parties, allowing them to compete on fair terms. Otherwise, all the while political power is geared inertially into the institutions, wealth and influence will continue to gravitate towards wealth and influence. The 'system' works to further the accretion of power to the already powerful, which in many areas means the accretion of land to the already landed.

It is the lack of any right to land which condemns hundreds of millions of people throughout the world to a life of material and spiritual poverty. Our affluent aspirations toward wisdom and joy cannot be held apart from these people's abject aspirations toward the most basic of livelihoods. It is impossible to lead a life of dignity when there is still so much appalling iniquity in the world. To be landless in such a world is to be powerless. And to be powerless is to feel a soul-withering sense of hopelessness. It is vital, therefore, that power be devolved downwards as well as upwards, from the institutions of the world to the people of the world. Through authentic democracy and authentic education, we must hope that more and more people will find the power to be able to help themselves, to bring more understanding to bear upon their problems and therefore be able to exert more control over their own lives. To take the most crucial example, we must hope that more and more women find the power and understanding to be able to choose for themselves to limit the size of their families.

The scale of the problems that face us, most particularly in respect of the exponential increase of the human population, is nothing less than frightening. But we must not be terrified into resignation. We cannot afford to be paralysed by the size of our task. We must ensure that the hopes and dreams of ordinary people come to prevail over the ambitions and power games of the nation states of this world. The Green agenda is now on the table. And politicians are actually beginning to take notice. There is a great deal that is encouraging. It is up to each one of us to keep the pressure up. We must not allow them to be complacent. And do not make the feeble excuse that your little bit of pressure won't make any difference. As I hope to have shown, your own superrational activity makes *all* the difference. Your participation is absolutely vital. We *must* dare to dream. Every letter you write, every pressure group subscription you pay, every conversation you hold, every kind of little protest that you can think to make; all will help foster the sanity which is the seed of our survival.

The realization of my *Earthdream* stands a long way off. Where are we to start? On a global scale, there can be little doubt about our first priority: the protection of our rainforests. This is an issue which has recently become the subject of a good deal of Western political rhetoric. The words are certainly a start, but we now require action. We need to press our politicians to *do* something, and urgently. Fine speeches are simply not enough. More than any other problem, time will not wait on this one. Here is James Lovelock, maverick scientist and author of *Gaia: a new look at life on Earth*, in an extract from his 1988 Schumacher lecture:

> The humid tropics are both a habitat for humans and a physiologically significant ecosystem. That habitat is being removed at a ruthless pace. Yet in the first world we try to justify the preservation of tropical forests on the feeble grounds that they are the home of rare species of plants and animals, even of plants containing drugs that could cure

cancer. They may do. They may even be slightly useful in removing carbon dioxide from the air. But they do much more than this. Through their capacity to evaporate vast volumes of water vapour, and of gases and particles that assist the formation of clouds, the forests serve to keep their region cool and moist by wearing a sunshade of white reflecting clouds and by bringing the rain that sustains them. Every year we burn away an area of forest equal to that of Britain and often replace it with crude cattle farms. Unlike farms here in the temperate regions such farms rapidly become desert, more trees are felled and the awful process of burning away the skin of the Earth goes on. We do not seem to realise that once more than 70 to 80 per cent of a tropical forest is destroyed the remainder can no longer sustain the climate and the whole ecosystem collapses. By the year 2000, at the present rate of clearance, we shall have removed 65 per cent of the forests of the humid tropics. After that it will not be long before they vanish, leaving the billion poor of those regions without support in a vast global desert. This is a threat greater in scale than a major nuclear war. Imagine the human suffering, the refugees, the guilt and the political consequences of such an event. And it will happen at a time when we in the first world are battling with the surprises and disasters of the greenhouse effect, intensified by the extra heating from the forest clearance. We may be in no position to help.[4]

I picture the Earth as a raft, afloat in a great black sea. For too long we have hacked and torn away at the timbers of our raft to fuel our insatiable appetite for 'things'. The superstructure is breaking up. Holes are appearing. Our raft is beginning to leak. Water is coming in. We clearly need to change our pattern of living. We have to start conserving the essential timbers that form the very body of our world. That world is not infinitely exploitable. We cannot carry on destroying the fabric of our own life-support system. It happens that the solution is quite straightforward, and really not very painful. We just have to accept the need to lead a simpler, more creative, less consumptive way of life. And a more committed way of life. We each need to spend just a little of our time each day baling the water out from our raft. This is our global responsibility.

But we are not yet taking this responsibility upon ourselves. We are continuing to strip the superstructure of our raft to fuel our material progress, and at an ever faster rate. And very few people are bothering to help bale. We somehow persuade ourselves that our individual effort will make so little difference as to be worthless. There is so much water, and our individual buckets are so small. Baling would be a waste of time. Wouldn't it? We decide that we have more important things to do with our time. Like watching television. And so our raft carries on letting in ever more water, through ever more holes.

The analogy is terrifyingly close to the reality. The seas *are* rising on us. Our great cities are threatened with a submarinal future. Does this bring us to our senses? No. We are too inured to apathy, too entrenched in our traditional ways, too snugly embraced in the comfort of our boredom, too much in love with our material vanities. With the water around our feet, and rising, we are still to be found hacking and tearing away at the timbers of our raft. The very real fear is that we will wake from our insanity too late. We will open our eyes to find the water around our necks. Our raft will be sinking. Or perhaps the water will only be around our knees. The holes will still be irreparable, but we might have a chance of survival—if we want to survive. Survival on those terms wouldn't be much fun. We would be required to spend the best part of each day just baling to keep the raft afloat, and you can imagine all the squabbles that would break out over that. It's a wretched prospect.

Yet, as things stand today, this is just the kind of dystopian future that we are going to be leaving for our children. I can hear their voices—angry voices, resentful voices. 'They were insane,' I hear them cry. 'They knew so much; how come they understood so little, how come they had so little vision?' I hear voices of despair. 'How could they have sat by and let this happen?' I hear voices of hatred. 'How could they have raped their own mother?'

Epilogue

It was always my intention that I would finish this work, just as I had started it, on a personal note. But when it came to the moment, I found myself losing my nerve. I began to get caught up with the thought that I should instead aim for some kind of intellectual denouement, to finally tie together the various strands of argument that have run through the book. However, after a few weeks spent fruitlessly writing around in circles, I have been forced to face up to what I really knew all along—that a book such as this can have no denouement. Its existential vectors will never be resolved in an analytical way. Its loose threads will not be tied up in a neat synoptic knot. And so, I had to hold my nerve.

This book has had a gestation period of some five years. As I write this Epilogue, it is June 1990. The events of the Prologue took place in July 1985. The trail joining the two has not followed the linear course that is apparent here. Indeed, it has taken the very opposite kind of course—frustratingly circuitous, often leading me back upon myself, and back upon myself again. It has often been rugged too, with signposts few and far between. I have had to survive several periods of crisis where I seemed to lose my way completely. At such times, struggling to recover the trail, I felt that I was never going to be able to put my thoughts and intuitions into words—let alone get them into print. Even if I could express myself well enough, would I ever be able to find someone willing to publish a book as idiosyncratic in style and as difficult to categorize as *Earthdream* was clearly going to be? The easy and the sensible thing to do, would have been to have given the project away and returned to pick up my career in computing. Was it not the most irresponsible folly to cast all my material security aside for something so intangible and uncertain? By most estimations, it almost certainly was, but I never felt as though there was any other choice open to me. Personal responsibility did not enter into the decision equation. I owed my allegiance to a greater sense of responsibility. It was somehow as if I had become involved in a collaboration with Fate.

Looking back, I can see now that, in part, this book has evolved as my answer to a question that not one of us can truly turn away from—the question of the nature of our religious belief. The question of where to find our source of value. In a way, I have tried to write the book that I would have liked to find when I first tried seriously to answer this question for myself, in the aftermath of the events that I described in the Prologue. Writing *Earthdream* has proved to be a sacramental act. It has enabled me to give some kind of form to my spiritual reality. However, it would be dishonest not to admit that it has also been just as much an act of redemption. In retrospect, I can see that my desire to write has also been fuelled by the need to redeem myself, not in the eyes of any God, but in the eyes of my children-to-be, and my grand-children-to-be.

Through my words, I think I was hoping that I could exonerate myself from blame for today's wanton ecological neglect. 'Look,' I could plead, 'you can condemn my generation, but you cannot condemn me. I spoke up. Here is the evidence.' But I fear now that the anger of future generations will not be placated so easily. 'Okay,' I hear them reply, 'you wrote a wonderful book, but what did you really *do*? Is it not true that writing your book simply enabled you to rest on your laurels, allowing you the luxury of actually *doing* very little?' I have come to understand that the completion of this book does not mark the end of my responsibility. It marks the beginning. Up until now, I have been able to excuse myself from responsibility because I have been writing. I have been too busy. In this way, the book has been my cover for five years. But of course it is about to be removed. And that is just why I have found myself procrastinating so much over this personal postscript. It hasn't been easy to admit that I am apprehensive about the prospect of my cover being blown. I think I'm going to feel very exposed.

In order to restore some modicum of financial stability to my life, I have at last returned to my career as a software analyst, developing mathematical models for a company which produces decision and planning support systems. After the long break, I am enjoying this involvement with computers again, getting up to date with all the amazing new software tools that have come upon the scene in the last few years. I am also enjoying the interaction which comes from working as a member of a team. I am even enjoying my thirty-minutes commuting on the train in the morning and evening, an hour of enforced stillness where I can completely lose myself in a book. And I am certainly enjoying the receipt of a regular salary cheque. Indeed, I fear that I am enjoying it too much. Already, I can feel myself becoming used to the regular routine of work. It seems that the more time I spend with my mind in the analytical gear that is required for my job, the less easy it is to make the shift back into intuition. Somehow, I don't feel as closely bonded to the Earth as I did just a month ago. The empathy is not so strong. It is as if I am being enveloped by an ever-thickening membrane, a desensitizing skin which is protecting my conscience from being touched too deeply by the greater concerns of the world. My evenings don't seem to give me enough time to dissolve that membrane away. Increasingly, I find that I am just too tired and too rushed even to want to try to dissolve it away.

I haven't yet become apathetic. I haven't yet become complacent. But I am beginning to see how easy it would be to let my will slip and succumb to the kind of spiritually devouring world-weariness that is our modern addiction to boredom. And it frightens me, because I can extrapolate this tendency to drift into mindless acquiescence across the entire population of the industrialized world. We are too exhausted working for our own personal dreams of security and success to be able to share in any collective dream—to share in the kind of greater dream, the global dream, that would allow us to envisage a sustainable presence upon our planet, that would allow humanity to become integrated as a *people*, living with wisdom and dignity upon the Earth. Sadly, as we are, without such a dream, we remain atomized as a collection of competing individuals within competing societies, cravenly content to let the Earth drift inexorably toward decay.

Corrupted by our Cartesian perception of reality, our dreamlessness is a result of our lack of a common symbol to express our common spirituality. Through our social structures, we have virtually nullified the authentic religious dimension to life. Our scientific self-definition contains no recognition of the reality of the religious spirit that potentially inspires us all. Indeed, as I have tried to show, it is this crisis of self-perception that lies at the heart of our ecological crisis. In the short term, a discerning Green consumerism will certainly help by placing pressure on industry, and a Greener system of economics will certainly help by taxing polluters into reform and encouraging conservation, but ultimately this kind of skin-deep Greening of our awareness is not going to be enough. The realization of authentic progress will require a soul-deep Greening of our perception. We are going to have to be Green to our spiritual core. Human survival is dependent on the Earth being adopted as the common symbol of our spirituality—the sacral centre of a new kind of religion, a new kind of mythology, a whole new way of seeing and thinking and living which embodies the marriage of reason and intuition.

With the writing of this Epilogue, I have reached the end of a long journey—only to find that I am being called forth to begin an even longer one. As I have already suggested, it is not a prospect that I feel altogether enthusiastic about right now. I feel that I need to recover from the rigours of this project before I can sign up for another long voyage. The temptation is to remove my name from Destiny's muster-roll and quietly consign myself to a secure and comfortable anonymity. But I know that I will not be able to betray my dream so easily. Albeit a little reluctantly, I find myself finally compelled to pledge here my commitment to a future collaboration with Fate. At present, I am unsure of the form that this collaboration will take, just as I am also unsure of the timing. For now, I think I am due a little bit of leave in port, to rest, to refit, to restock on essential gumption, to await the right tide to take me out to sea again. I want some time purely for myself, and for my wife. I want to spend some time in the mountains. I want to spend some time reading novels. These are simple pleasures that I've had to deny myself this last year. Before too long, though, I know that orders to sail will arrive from somewhere within my soul. I know that I will be obliged to find further ways of sowing the seeds of my *Earthdream*. This is just too important a dream to be allowed to wither away.

If I give free rein to my imagination, I see *Earthdream* becoming a word which grows in meaning to encapsulate the whole idea of a Green mythology. I see *Earthdream* becoming a vehicle for the promotion of the earth as the common symbol of human spirituality. I see *Earthdream* becoming a movement, a collective vision of an entire people, working to realize the cultural transformation outlined in this book. I give my dreams room to expand in this way, and find myself awed by the possibilities—because I know, in my heart, that with our full commitment these possibilities could become actualities. It is within our power, the power of people like you and me, to make my *Earthdream* a reality. Aware of that truth, we have no choice but to dream big dreams. To have no dream is to have no vision, and to have no vision is to have no future.

References

Chapter 1:   The Mythology of Science

1. Fritjof Capra, *The Turning Point*. London: Fontana Flamingo, 1985, p. 23.
2. Erwin Schrödinger, *Mind and Matter*, Cambridge: Cambridge University Press, 1958, p. 65.
3. Werner Heisenberg, *Physics and Philosophy*. London: Allen and Unwin, 1959, p. 57.
4. Erwin Schrödinger, 'Are there Quantum Jumps', *The British Journal for the Philosophy of Science*, Vol. 3 (1952), pp. 109–10 (as quoted in Ilya Prigogine & Isabelle Stengers, *Order out of Chaos*. London: Fontana Flamingo, 1985, p. 18).

Chapter 2:   The Physics of Participation

1. Morris Berman, *The Reenchantment of the World*. New York: Bantam, 1988, p. 3.
2. Arthur Eddington, *The Nature of the Physical World*. Cambridge: Cambridge University Press, 1928, pp. xvi–xvii.
3. Paul Davies, *The Cosmic Blueprint*. London: Unwin Hyman, 1989, p. 168.
4. Morris Berman, *The Reenchantment of the World*. op. cit., p. 137.

Chapter 3:   Exorcising the Ghost of Cartesianism

1. Ken Wilber, *Quantum Questions*. Boston: Shambhala, 1985, p. 10.
2. Arthur Eddington, *Science and the Unseen World*, New York: Macmillan, 1929 (as quoted in Ken Wilber, *Quantum Questions*. op. cit., p. 10).
3. Ken Wilber, *No Boundary*, Boston: Shambhala, 1985, p. 50.
4. Ibid., p. 52.
5. Arthur Eddington, *The Nature of the Physical World*. op. cit., pp. xvi–xvii.

6. Erwin Schrödinger, *My View of the World*, trans. Cecily Hastings. Cambridge: Cambridge University Press, 1964, p. 18.
7. Ibid., pp. 20–2.
8. Morris Berman, *The Reenchantment of the World*. op. cit., p. 176.
9. Colin Wilson, *Beyond the Outsider*. London: Pan, 1966, p. 75.
10. Gregory Bateson, *Steps to an Ecology of Mind*. New York: Ballantine, 1972, p. 461.
11. Morris Berman, *The Reenchantment of the World*. op. cit., p. 178.

Chapter 4:   The Mythology of Religion

1. Joseph Campbell, *Myths to Live By*. London: Granada Paladin, 1985, p. 19.
2. Ibid., p. 19.
3. Aldous Huxley, *The Perennial Philosophy*. London: Triad Grafton, 1985, p. 9.
4. Alan Watts, *The Book: On the Taboo Against Knowing Who You Are*. New York: Vintage, 1972, p. 12.
5. Ibid., p. 13.
6. Ibid., p. 10.
7. Kahlil Gibran, *The Prophet*. London: Heinemann Pan, 1980, p. 36.
8. Joseph Campell, *Myths to Live By*. op. cit., p. 19.
9. Joseph Campell, *The Power of Myth*. New York: Doubleday, 1988, p. 207.
10. Joseph Campell, *Myths to Live By*. op. cit., p. 10.

Chapter 5:   The *Tao* of Wisdom

1. Marilyn Ferguson, *The Aquarian Conspiracy*. London: Grafton Paladin, 1982, p. 119.
2. Lao Tzu, *Tao Te Ching*, trans. Stephen Mitchell. New York. Harper & Row, 1988, Ch. 1.
3. Lao Tzu, *Tao Te Ching*, trans. Gia-fu Feng & Jane English. London: Wildwood House, 1973, Ch. 14.
4. Lao Tzu, *Tao Te Ching*, trans. Stephen Mitchell. op. cit., Ch. 22.
5. Ibid., Ch. 47.
6. Ibid., Ch. 21.
7. Colin Wilson, *Beyond the Outsider*, op. cit., p. 82.
8. Ibid., p. 31.
9. Robert M. Pirsig, *Zen and the Art of Motorcycle Maintenance*. London: Corgi, 1976, p. 296.
10. Ibid., p. 297.
11. Colin Wilson, *The Occult*. London: Hodder & Stoughton, 1971, p. 136.

12. Colin Wilson, *Frankenstein's Castle*. Bath: Ashgrove Press, 1980, p. 115.
13. Colin Wilson, *New Pathways in Psychology*. London: Victor Gollancz, 1972, p. 31.
14 Susan Griffin, *The Schumacher Lectures* Vol. 2. London: Sphere Abacus, 1986, pp. 175–6.

Chapter 6: Being: An Existential Rainbow

1. Marilyn Ferguson, *The Aquarian Conspiracy*. op. cit., p. 120.
2. Abraham Maslow, *Toward a Psychology of Being*. New York: Van Nostrand Reinhold, 1968, pp. 160–1.
3. Rollo May, *The Courage to Create*. New York: Bantam, 1976, pp. 4–5.
4. Alan Watts, *The Meaning of Happiness*. London: Rider, 1978, p. 180.
5. Ken Wilber, *No Boundary*. op. cit., pp. 80–1.
6. Erich Fromm, *You Shall Be As Gods*. New York: Ballantine, 1983, p. 48.
7. Sören Kierkegaard, *Sickness unto Death*, trans. Walter Lowrie. Princeton: Princeton University Press, 1941 (as quoted in Rollo May, *The Meaning of Anxiety*. New York: Washington Square Press, 1979, p. 356).
8. Abraham Maslow, *Toward a Psychology of Being*. op. cit., p. 46.
9. Rollo May, *The Meaning of Anxiety*. op. cit., pp. 74–5.
10. Ken Wilber, *No Boundary*. op. cit., p. 92.
11. Joan Borysenko, *Minding the Body, Mending the Mind*. London: Bantam, 1988, p. 95.
12. Rollo May, *The Courage to Create*. op. cit., p. 135.

Chapter 7: The Mythology of Progress

1. Laurens van der Post, *The Heart of the Hunter*. London: Penguin, 1965, pp. 128–9.
2. Ibid., p. 129.
3. D.H. Lawrence, 'Etruscan Cypresses', *The Complete Poems of D.H. Lawrence* Vol. 1. London: Heinemann, 1964, p. 298.
4. James P. Carse, *Finite and Infinite Games*. London: Penguin, 1987, p. 3.
5. Ibid., p. 31.
6. Ibid., p. 31.
7. E.F. Schumacher, *Small is Beautiful*. London: Sphere Abacus, 1974, p. 124.
8. Ibid., p. 124.

---

9. Ibid., p. 124.
10. Ibid., p. 124.
11. Erich Fromm, *Beyond the Chains of Illusion*. London: Sphere Abacus, 1980, p. 119.
12. James P. Carse, *Finite and Infinite Games*. op. cit., p. 26.

Chapter 8: Education as Enchantment

1. James P. Carse, *Finite and Infinite Games*. op. cit., p. 19.
2. Idries Shah, *Thinkers of the East*. London: Penguin, 1974, p. 137.
3. Ibid., p. 82.
4. James P. Carse, *Finite and Infinite Games*. op. cit., p. 109.
5. Ibid., p. 96.
6. Ibid., p. 105.
7. Ibid., p. 37.
8. Ibid., p. 121.

Chapter 9: Evolution as Involution

1. Kathleen Raine, 'Earth's Children', *Resurgence*, 135 (1989), p. 4.
2. Thomas Berry, *The Dream of the Earth*. San Francisco: Sierra Club, 1988, p. 10.
3. Ibid., p. 132.
4. Ibid., p. 198.
5. Murray Eden, 'Monograph No. 5', *Mathematical Challenges to the neo-Darwinian Interpretation of Evolution*, eds. P.S. Moorhead & M.M. Kaplan. Philadelphia: Wistar University Press, 1967 (as quoted in Francis Hitching, *The Neck of the Giraffe*. London: Pan, 1982, p. 83).
6. Albert Einstein, quoted in J. Gowan, *Trance, Art, and Creativity*. California, 1975 (as quoted in Ken Wilber, *Up from Eden*. London: Routledge & Kegan Paul, 1983, p. 4).

Chapter 10: The Mythology of Survival

1. Douglas Hofstadter, *Metamagical Themas*. London: Penguin, 1986, p. 757.
2. Milan Kundera, *The Unbearable Lightness of Being*, trans. Michael Henry Heim. London: Faber & Faber, 1985, p. 5.
3. Jonathon Porritt, 'Green Consumerism', *Resurgence*, 137 (1989), p. 10.
4. James Lovelock, 'Stand Up For Gaia', *Resurgence*, 132 (1989), pp. 9–10.